Zhiyong Zhang and Ke-Hai Yuan

Practical Statistical Power Analysis

using WebPower and R

ISDSA Press · Granger, IN

PUBLISHED BY ISDSA PRESS · GRANGER, IN

WWW.ISDSA.ORG

ISBN-13 (pbk): 978-1-946728-02-9
ISBN-13 (electronic): 978-1-946728-00-5

Library of Congress Cataloging-in-Publication Data
Personal name: Zhang, Zhiyong (Professor), author
Main title: Practical Statistical Power Analysis using WebPower and R / Zhiyong Zhang and Ke-Hai Yuan, University of Notre Dame, Notre Dame, Indiana, USA
Published/Produced: ISDSA Press
LCCN: 2018901582
LCCN Permalink: https://lccn.loc.gov/2018901582

First printing, June, 2018

Preface

In designing a survey or an experiment, statistical power analysis is often used to determine the sample size – the number of participants or subjects. This ensures that one would detect an existing effect when conducting a hypothesis testing with the collected data. Since a statistical power analysis is often specific to a particular statistical test, the procedure for power analysis has to be developed for each test accordingly. For example, the method for a t-test is different from that for a regression in general.

Not surprisingly, statistical power analysis is a popular research area in statistics. The work by Cohen (1988) is probably the most influential piece in the area. Availability of software also makes statistical power analysis possible. For example, G*Power (Faul et al., 2009) is a popular, free, stand-alone program for a large number of types of power analysis. Similarly, commercial software such as SAS and Stata also includes procedures or routines for power analysis.

Supported by a grant from the Institute of Education Sciences (R305D140037), we have conducted a study trying to develop general methods and software tools for statistical power analysis. First, we studied the existing methods for power analysis and tried to incorporate them into our software development. Second, we discussed a general Monte Carlo based method for statistical power analysis, which seemed to become more and more popular in the literature (e.g., Muthén & Muthén, 2002; Zhang, 2014). Third, we developed a new double Monte Carlo method to deal with missing data and non-normal data in planning sample sizes.

The contents of the book and WebPower do not necessarily represent the policy of the Department of Education, and you should not assume endorsement by the Federal Government.

To help the adoption of the methods, we also developed software in several formats.

- A Web application WebPower was developed and is currently online so that one can conduct statistical power analysis freely online. Therefore, statistical power analysis can be conducted within a Web browser directly on a desktop, laptop or even a smartphone. There is generally no need to install any software for carrying out power analysis.

https://webpower.psychstat.org

2

- An R package WebPower is made available on CRAN. Therefore, one can conduct power analysis in the popular statistical software R. As an open-source project, interested researchers are welcome to contribute to the development of the package on GitHub.

  https://CRAN.R-project.org/package=WebPower

  https://github.com/johnnyzhz/WebPower

- An Android App WebPower is now published on Play store. Anyone can install the app on their Android devices and conduct the analysis on a phone or tablet.

This book serves as both a manual for the software we developed and a technical report for the methods underlying the software. By using the book, readers will immediately learn how to conduct statistical power analysis through well-designed examples. For each power analysis method, we also tried our best to provide the underlying technical details. This helps interested readers understand the method underlying each power analysis procedure.

The book consists of three parts, after the first introductory chapter. Chapter 1 provides a general introduction to statistical power analysis. Part I consists of Chapters 2 to 15. Each chapter in this part focuses on a particular type of power analysis. For example, Chapter 2 is about power analysis for testing correlation coefficients. Chapter 15 discusses two methods for conducting a power analysis for structural equation modeling. Part II consists of Chapters 16 to 18. In this part, we show how to conduct regular Monte Carlo based power analysis for the t-test, general SEM and mediation analysis, and latent score analysis. Part III introduces a general path diagram based method for statistical power analysis. Both the regular Monte Carlo method and the double Monte Carlo method are discussed.

The study and the book are not possible without the support from the Institute of Education Sciences (IES; R305D140037). The Center for Research Computing at the University of Notre Dame hosts the servers running the WebPower Web application and provides technical support.

https://crc.nd.edu/

Our study was also supported by many individuals including our colleagues in the Department of Psychology at the University of Notre Dame. Many graduate students have also contributed to the writing of the book including Meghan Cain, Han Du, Ge Jiang, Haiyan Liu, Agung Santoso, Lin Xing, and Miao Yang. Dr. Yujiao Mai assisted in the development of the software and contributed to several chapters of the book. Their names and current affiliations are listed within each chapter. Last but not least, we would like to thank our program officer at IES – Dr. Phill Gagné. Dr. Gagné has been always supportive and patient over the whole course of the project.

Zhiyong Zhang and Ke-Hai Yuan
Notre Dame, IN, 2018

Contents

4

1   *Introduction to Statistical Power Analysis*

Zhiyong Zhang and Ke-Hai Yuan
Department of Psychology
University of Notre Dame

Performing statistical power analysis and sample size estimation is an important aspect of experimental design. Journals published by both American Educational Research Association and American Psychological Association have particularly emphasized the importance of statistical power analysis (cf., Peng et al., 2012). Without power analysis, the sample size may be too large or too small. If a sample size is too small, an experiment will lack the precision to provide reliable answers to the questions under investigation. Then, the validity of the statistical conclusions from the research is endangered (e.g., Cohen, 1988; Hedges & Rhoads, 2010; Shadish et al., 2002). If a sample size is too large, time and resources will be wasted, often for minimal gain. Statistical power analysis and sample size planning allow us to decide how large a sample size is needed to enable statistical judgments that are accurate and reliable and how likely a statistical test is going to detect effects of a given size in a particular situation.

1.1   *What Is Statistical Power?*

The power of a statistical test is the probability that the test will reject a false null hypothesis (i.e. that it will not make a type II error). Given the null hypothesis H_0 and an alternative hypothesis H_1, we can define power in the following way. First, the type I error is the probability to incorrectly reject the null hypothesis. Therefore

$$\text{Type I error} = \Pr(\text{Reject } H_0 | H_0 \text{ is true}).$$

The type II error is the probability of failing to reject the null hypothesis while the alternative hypothesis is correct. That is

$$\text{Type II error} = \Pr(\text{Fail to reject } H_0 | H_1 \text{ is true}).$$

Statistical power is the probability of correctly rejecting the null hypothesis when the alternative hypothesis holds. That is:

$$\text{Power} = \Pr(\text{Reject } H_0 | H_1 \text{ is true}) = 1 - \text{Type II error}.$$

We can summarize these in the table below.

	Fail to reject H_0	Reject H_0
Null Hypothesis H_0 is true	Good	Type I error
Alternative Hypothesis H_1 is true	Type II error	Power

1.2 Factors Influencing Statistical Power

Statistical power depends on many factors. In particular, power nearly always depends on the following three factors: the statistical significance criterion (e.g., alpha level or type I error), the effect size and the sample size. In general, power increases with larger sample size, larger effect size, and larger alpha level.

1.2.1 Alpha level or significance level

A significance criterion is a value for us to determine whether the observed statistic is extreme enough so that we can declare that H_0 does not hold. The most commonly used criteria are alpha level, or type I error rate 0.05 (5%, 1 in 20), 0.01 (1%, 1 in 100), and 0.001 (0.1%, 1 in 1000). If the alpha level is set at 0.05, the probability of obtaining a significant effect when the null hypothesis is true should be around 0.05[1], and so on. The simplest way to increase the power of a test is to carry out a less conservative test by using a larger significance level. This increases the chance of obtaining a statistically significant result (rejecting the null hypothesis) when the null hypothesis is false. However, it also increases the risk of obtaining a statistically significant result when the null hypothesis is true. By default, a significance level of 0.05 is commonly used.

[1] In practice, the empirical probability will not be exactly 0.05. A value greater than 0.05 means the test would reject the null hypothesis more than it should, which might lead to inflated power.

1.2.2 Effect size

In power analysis, effect size is used to quantify the magnitude of the effect of interest in the population. An effect size can be a direct measure of the quantity of interest, or it can be a standardized measure that also accounts for the distribution of the population. For example, in an analysis comparing outcomes in treatment and control, the difference of outcome means $\mu_1 - \mu_2$ would be a direct measure of the effect size, whereas $(\mu_1 - \mu_2)/\sigma$ is a standardized effect size, where σ is the common standard deviation of the outcomes in the populations under the conditions of treatment and control. When the population distributions are known, a standardized effect size, along with the sample size and alpha level, will completely determine the power such as under the t-distribution or F-distribution. An unstandardized (direct) effect size alone will rarely be sufficient to determine the power, as it does not contain information about the variability in the measurements.

1.2.3   Sample size

Sample size is a primary factor that affects the power. Sample size is related to the amount of sampling error inherent in the result of a test. Other things being equal, effects are harder to detect in smaller samples. Increasing sample size directly increases the statistical power of a test. However, a large sample size would need more resources to achieve, which might not be feasible at all in practice.

1.2.4   Other factors

Many other factors can influence statistical power. First, increasing the reliability of data can increase power. The precision with which the data are measured influences statistical power. Consequently, power can often be boosted by reducing the measurement error in the data (Yuan & Bentler, 2006). A related concept is to improve the "reliability" of the measure being assessed (as in psychometric reliability).

Second, statistical power can also be improved by better design of an experiment or observational study. For example, in a two-sample testing situation with a given total sample size n, it is optimal to have equal numbers of observations from the two populations being compared (as long as the variances in the two populations are the same). In regression analysis and Analysis of Variance (ANOVA), there are extensive theories as well as practical strategies, for increasing the power based on optimally setting the values of the independent variables in the model.

Third, for longitudinal studies, power increases with the number of measurement occasions (Zhang & Liu, 2018; Zhang & Wang, 2009). Power may also be related to the time interval of measurements (Cain et al., 2018).

Fourth, missing data reduce effective sample size and thus statistical power (Zhang & Wang, 2009). Furthermore, different missing data patterns can have different power.

Last but not least, statistical power analysis is often based on the assumption that the collected data will be normally distributed. However, real data are often nonnormally distributed, due to data contamination or ordinal measurements. It has been found that non-normality of data will lead to decreased power especially when the normal based statistical methods are used for data analysis (Yuan et al., 2015; Zhang, 2014).

1.3   Conducting Power Analysis and Sample Size Planning

To ensure that a statistical test will have adequate power, we usually must perform special analyses by considering all possible factors prior to running the experiment to calculate how large a sample size n is required. Although there are no formal standards for power, most researchers assess the power using 0.80 as a standard for adequacy. This convention implies a four-to-one tradeoff between Type II error and Type I error.

We now use a simple example to illustrate how to calculate power and sample size. More complex power analysis can be conducted similarly.

Suppose a researcher is interested in whether training can improve mathematical ability. She can conduct a study to get the math test scores from a group of students before and after training. The null hypothesis here is that training does not improve math ability or the change is 0. She would like to have enough power to detect a change of one unit. Thus, with μ representing the population mean of change, the alternative hypothesis is that the change is 1:

$$H_0 : \mu = \mu_0 = 0$$

$$H_1 : \mu = \mu_1 = 1.$$

Based on the definition of power, we have

$$
\begin{aligned}
\text{Power} &= \Pr(\text{reject } H_0 | \mu = \mu_1) \\
&= \Pr(\text{change } (d) \text{ is larger than critical value under } H_0 | \mu = \mu_1) \\
&= \Pr(d > \mu_0 + c_{1-\alpha}\sigma/\sqrt{n} | \mu = \mu_1)
\end{aligned}
$$

where

- μ_0 is the population value of change under the null hypothesis.

- μ_1 is the population value of change under the alternative hypothesis.

- d is the observed change before and after training.

- σ is the population standard deviation under the null hypothesis.

- c_α is the critical value for a distribution, such as the standard normal distribution or t-distribution.

- n is the sample size.

Note that to calculate the power, we need to know $\mu_0, \mu_1, \sigma, c_\alpha$, the sample size n, and the distributions of d under both the null hypothesis and alternative hypothesis. Let us assume that $\alpha = .05$ and the distribution is normal with the same variance σ under both null and alternative hypothesis. Then the above power is

Calculate power

$$
\begin{aligned}
\text{Power} \ &= \ \Pr(d > \mu_0 + c_{1-\alpha}\sigma/\sqrt{n}|\mu = \mu_1) \\
&= \ \Pr(d > \mu_0 + c_{1-\alpha} \times \sigma/\sqrt{n}|\mu = \mu_1) \\
&= \ \Pr\left[\frac{d - \mu_1}{\sigma/\sqrt{n}} > -\frac{(\mu_1 - \mu_0)}{\sigma/\sqrt{n}} + c_{1-\alpha}|\mu = \mu_1\right], \quad (1.3.1) \\
&= \ 1 - \Phi\left[-\frac{(\mu_1 - \mu_0)}{\sigma/\sqrt{n}} + c_{1-\alpha}\right] \\
&= \ 1 - \Phi\left[-\frac{(\mu_1 - \mu_0)}{\sigma}\sqrt{n} + c_{1-\alpha}\right] \\
&= \ 1 - \Phi\left[-\frac{(\mu_1 - \mu_0)}{\sigma}\sqrt{n} + 1.645\right]. \quad\quad (1.3.2)
\end{aligned}
$$

where Φ is the cumulative distribution function of a standard normal random variable. Thus, power is related to sample size n, the significance level α, and the effect size $\delta = (\mu_1 - \mu_0)/\sigma$. Using the effect size δ, Power $= 1 - \Phi(-\delta\sqrt{n} + c_{1-\alpha})$. Since Φ is an increasing function, the power

- increases with the increase of the sample size n,

- increases with the increase of the effect size δ, and

- increases with the increase of the alpha level α because it decreases the critical value $c_{1-\alpha}$.

For example, if we assume $\sigma = 2$, then the effect size is $0.5 = (1-0)/2$. When the sample size is 100, the power from the above formulae is .999 by simply plugging in the numbers.

Using R:
```
> 1-pnorm(-.5*sqrt(100)+qnorm(.95))
[1] 0.9996
```

If we know the power, we can solve Equation 1.3.1 to get the corresponding sample size

Sample size planning

$$
n = \left[\frac{c_{1-\alpha} - \Phi^{-1}(1 - \text{Power})}{\delta}\right]^2.
$$

For example, when the power is 0.8, we can find a sample size of 25.

In many situations, one might have a good idea on how many subjects can be recruited in a study based on available resources, but may not have much information on the effect size. In this case, the

Using R:
```
> ((qnorm(1-.05)-qnorm(1-.8))/.5)^2
[1] 24.73
```

minimum detectable effect size can be calculated, which is the effect size needed to get significant results based on expected power. For the current example, we can obtain δ by

$$\delta = \frac{c_{1-\alpha} - \Phi^{-1}(1 - \text{Power})}{\sqrt{n}}.$$

Minimum detectable effect size
Using R:
```
> (qnorm(1-.05)-qnorm(1-.8))/sqrt(20)
[1] 0.55599
```

For example, if one can only collect data from 20 participants, to get a power 0.8 the effect size has to be at least 0.56.

Although it is very rare, one can also calculate the alpha level to achieve certain power with known sample size and effect size. In this case,

$$\alpha = 1 - \Phi\left[\delta\sqrt{n} + \Phi^{-1}(1 - \text{Power})\right].$$

For example, to get a power 0.8 with the sample size 20 and effect size 0.5, the alpha level has to be increased to 0.08.

Using R:
```
> 1-pnorm(.5*sqrt(20)+qnorm(1-.8))
[1] 0.081591
```

Generally speaking, if there is an analytical form for power calculation, any one unknown quantity can be quickly calculated when other quantities are known. In the above example, given any three of the four values – n, Power, δ and α – the fourth one can be calculated.

Practical power analysis can, and most likely will be, more complex than the example above. This book aims to provide a general tool for conducting power analysis for a variety of models used in practical research. However, even for more complex power analysis, the basic idea discussed here applies.

1.4   Retrospective or Post-hoc Power Analysis

Statistical power analysis or sample size planning generally should be conducted before the beginning of a study by assuming the population effect size is known. This kind of power analysis is also called prospective power analysis or *a priori* power analysis. The use of retrospective or post-hoc power analysis, conducting a power analysis after a non-significant hypothesis testing, is highly controversial.

Hoenig & Heisey (2001) argued that post-hoc power is a function of the p-value. For example, they showed that for a two-tailed Z test, if the p-value is 0.05, the post-hoc power would be 0.5. Therefore, once a p-value is calculated, there is no need to calculate post-hoc power.

Yuan & Maxwell (2005) more concretely showed that the post-hoc power is almost always a biased estimator of the true power. The bias can be positive or negative, and does not become smaller as the sample size increases. Thus, the calculation of post-hoc power does not provide valuable information about the study under investigation and is therefore not very useful.

In this book, all of the power analyses are prospective even though a power analysis might be conducted based on an estimated effective size from an empirical or existing study. In particular, we need to assume that the population effect size is at the given value regardless of how such a value is obtained. A sample size based on such an assumption will be obtained with the desired power. If the following study does not yield a significant result at the suggested sample size, then we may conclude that the real effect size is not as large as what we have assumed/estimated.

1.5   *Determination of Effect Size*

As we have shown earlier, in calculating a statistical power, the population effect size needs to be provided. In practice, the population effect size is rarely known and determining the population effect size is always a subjective process. Therefore, we suggest the following strategies concerning effect size when calculating power for planning a study.

First, if there are well-accepted benchmark effect sizes, they can be used in sample size planning. For example, for correlation it is well established that 0.1, 0.3, and 0.5 represent small, medium, and large effects (Cohen, 1988). If a researcher believes that the population effect is medium, then he or she can use 0.3 as the population effect size in the power calculation. If an effect between small and medium is expected, one can generate a power curve by considering different effect sizes within the interval [0.1, 0.3].

Second, an effect size can be estimated from a pilot study, published research, or other empirical sources. This is the sample effect size. In many scenarios, a sample effect size has been used as if it is the population effect size. However, extreme caution should be used here because (1) the sample effect size is influenced by sample size and the population distribution, (2) the sample effect size can be greater or smaller than the population effect size even in a random sample (Du & Wang, 2016), and (3) publication bias can easily cause bias in the sample effect size even based on meta-analysis (Du et al., 2017a). Sampling errors cannot be avoided even when the estimated effect size is unbiased, and any such error will be multiplied by N (the sample size) or the square root of N in the power function (Yuan & Maxwell, 2005).

Third, there are situations where it is extremely difficult to come up with an effect size. In this case, one can calculate the minimum effect size to achieve certain power as we discussed earlier about the minimum detectable effect size.

Part I

Statistical Power Analysis
based on Analytical Results

2 | *Statistical Power Analysis for Correlation*

Lin Xing
Department of Psychology
University of Notre Dame

Correlation measures the direction and strength of the linear relationship between two variables. The most widely used correlation coefficient is the Pearson product moment correlation coefficient. Sample correlation coefficients (r) estimate the effect in the population (ρ). Values of the correlation coefficient are always between -1 and +1. A coefficient of +1 indicates that two variables are perfectly and positively linearly related, a coefficient of -1 indicates that two variables are perfectly and negatively linearly related, and a coefficient of 0 indicates that there is no linear relationship. This chapter focuses on how to calculate power for testing the Pearson product moment correlation between two continuous variables.

2.1 | *How to Conduct Power Analysis for Correlation*

The primary software interface for power analysis of correlation is shown in Figure 2.1.1. Within the interface, a user can supply different parameter values and select different options for power analysis. Among the four parameters, *Sample size, Correlation, Significance level,* and *Power,* one and only one can be left blank.

URL: http://w.psychstat.org/correlation

- The *Sample size* is the total number of subjects or participants in a study. For example, if there are 100 participants in a study, the sample size is 100. Multiple sample sizes can be provided in two ways to calculate power for each sample size. First, multiple sample sizes can be separated by spaces. For example, 100 150 200 will calculate power for the sample sizes 100, 150, and 200. Second, a sequence of sample sizes can be generated using the method *s:e:i* with *s* denoting the starting sample size, *e* the ending sample size, and *i* the interval. Note that the values are separated by a colon ":".

Correlation Coefficient

Parameters (Help)

Sample size	100
Correlation	0.1
# of vars partialed out	0
Significance level	0.05
Power	
H1	Two sided ⬍
Power curve	No power curve ⬍
Note	Power for correlation

Calculate

For example, 100:150:10 will calculate power for the sample sizes 100, 110, 120, 130, 140, and 150. The default sample size, as shown in Figure 2.1.1, is 100.

- The *Correlation* specifies the population correlation coefficient (ρ). It could be input directly based on the theory or be left blank and calculated given sample size, power and significance level. Multiple correlation coefficients or a sequence of correlation coefficients can be provided using the same way method as sample size. By default, the value is 0.1.

- The *Number of variables partialed out* could be used to conduct partial correlation analysis. For example, if there is one variable partialed out, the application will calculate power for the partial correlation of two variables while taking away the effect of the third variable. The default value is 0, which is used to calculate the power for correlation analysis without partialing out the effect of any other variables.

- The *Significance level* (Type I error rate or alpha level) for power calculation is required, but usually is set at the default value of 0.05.

- The *Power* specifies the desired statistical power, usually 0.8, or can be left blank to have power calculated given sample size, effect size, and significance level.

- The *H1* can be specified as "Two-sided", "Less", or "Greater" based on the alternative hypothesis.

- In addition to the required input, a plot of the power curve can also be requested if multiple sample sizes are provided.

- A *Note* (less than 200 characters) can also be used to provide basic information of the analysis for future reference.

2.1.1   *Examples*

To illustrate how to use the software WebPower for power analysis for correlation, we provide several examples.

Suppose a student wants to study the relationship between stress and health. Based on her prior knowledge, she expects the two variables to be correlated with a correlation coefficient of 0.3. If she plans to measure the stress and health of 50 participants, what is the power to obtain a significant correlation?

The input and output for calculating power for this study are given in Figure 2.1.2. The total number of participants, 50, and the expected correlation, 0.3, are input into their corresponding fields. The default number of variables partialed out, 0, is used. The default significance level 0.05 is used although one can change it to a different value. The field for *Power* is left blank because it will be calculated. The *H1* is "Two-sided" corresponding to the null hypothesis that the correlation is 0 and the alternative hypothesis that the correlation is not 0. A simple note "Power for correlation" is also added in the *Note* field. By clicking the "**Calculate**" button, the statistical power is given in the output immediately. For the current design, the power is 0.5729.

A power curve is a line plot of the statistical power as sample size varies. In Example 2.1.1, the power is 0.5729 with the sample size 50. What is the power for other sample sizes? One can investigate the power of different sample sizes and plot a power curve.

The input and output for calculating power for the study in Example 2.1.1 with a sequence of sample sizes from 50 to 100 with an interval of 10 are given in Figure 2.1.3. In the *Sample size* field, the input is 50:100:10, and we also choose "*Show power curve*" for the *Power curve* field in the input. In the output, the power for each sample size from 50 to 100 with the interval 10 is listed, and the power curve is displayed at the bottom of the output as shown in Figure 2.1.4. The power curve can be used for interpolation. For example, to get a power 0.8, a sample size of 85 is needed.

In practice, a power 0.8 is often desired. Given the desired power, the sample size can also be calculated as shown in Figure 2.1.5. In this situation, the *Sample size* field is left blank while in the *Power* field, the value 0.8 is input. In the output, we can see that the sample size 84, rounded to the nearest integer, is needed to obtain the power 0.8.

Example 2.1.1: Calculate power given sample size and effect size

Example 2.1.2: Power curve

Example 2.1.3: Calculate sample size given power and effect size

Correlation Coefficient

Figure 2.1.2: Input and output for calculating power for correlation in Example 2.1.1.

Parameters (Help)

Sample size   50

Correlation   0.3

# of vars partialed out   0

Significance level   0.05

Power

H1   Two sided ⬍

Power curve   No power curve ⬍

Note   Power for correlation

Calculate

Output

```
Power for correlation

    n    r alpha  power
   50 0.3  0.05 0.5729


URL: http://psychstat.org/correlation
```

Example 2.1.4: Minimum detectable effect size

Sometimes, it may be difficult to know the effect size in advance, while the sample size and power can be easily pre-determined. In this situation, one can calculate the minimum effect size required to obtain a significant results with the desired power. The input and output for this calculation are given in Figure 2.1.6. The *Correlation* field is left blank. The desired power, 0.8, and the total number of participants, 50, are input into their corresponding fields. The output shows that a correlation (effect size) of 0.3838 is needed to obtain a power 0.8. This means that the population correlation has to be at least 0.3838 to obtain a power 0.8 with the sample size 50.

Example 2.1.5: Calculate power for a partial correlation

The time that people spend on doing exercise each week may affect the correlation between stress and health. Therefore, the student might want to investigate the correlation between stress and health after controlling for the effect of time spent exercising. Suppose the correlation after partialing out the effect of time is 0.24. In this example, the *Number of variables partialed out* should be supplied with "1" as shown in Figure 2.1.7. In the output, the power for this partial correlation is 0.3889 with

Correlation Coefficient

Parameters (Help)

Sample size	50:100:10
Correlation	0.3
# of vars partialed out	0
Significance level	0.05
Power	
H1	Two sided ⬍
Power curve	Show power curve ⬍
Note	Power for correlation

Calculate

Output

```
Power for correlation

    n    r alpha  power
   50 0.3  0.05 0.5729
   60 0.3  0.05 0.6542
   70 0.3  0.05 0.7230
   80 0.3  0.05 0.7803
   90 0.3  0.05 0.8272
  100 0.3  0.05 0.8652


URL: http://psychstat.org/correlation
```

a sample size 50.

2.2   *Effect Size for Correlation*

The Pearson product moment correlation coefficient itself has been used as a standardized effect size (Cohen, 1988). It is defined as

$$r = \frac{\sum_{i=1}^{n}(x_i - \bar{x})(y_i - \bar{y})}{\sqrt{\sum_{i=1}^{n}(x_i - \bar{x})^2}\sqrt{\sum_{i=1}^{n}(y_i - \bar{y})^2}}.$$

Let r_{AB}, r_{AC}, r_{BC} be the correlation between A and B, A and C, and B and C, respectively. The partial correlation coefficient is defined as

$$r_{AB|C} = \frac{r_{AB} - r_{AC}r_{BC}}{\sqrt{(1 - r_{AC}^2)(1 - r_{BC}^2)}},$$

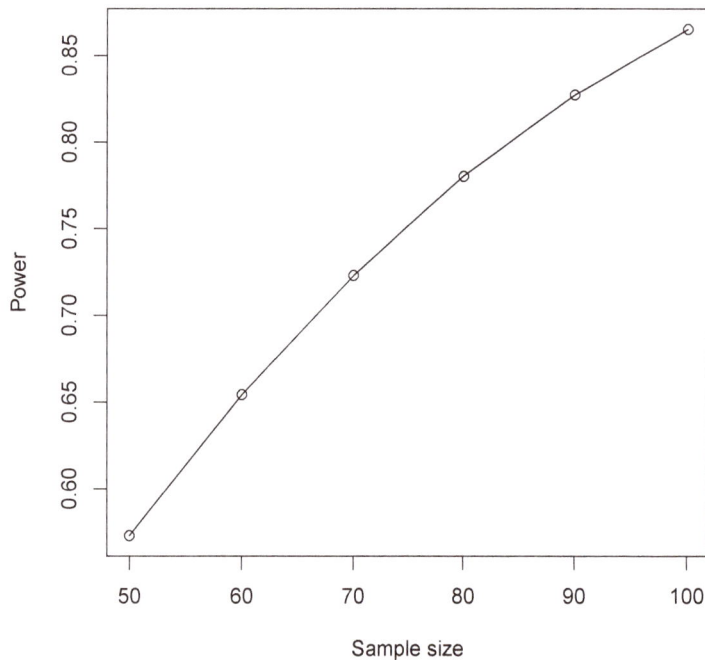

which is the correlation of A and B after controlling for the effect of C. The correlation coefficient is a standardized metric, and effects reported in the form of r can be directly compared across samples. According to Cohen (1988), a correlation coefficient of .10 is considered to represent a weak or small association; a correlation coefficient of .30 is considered as a moderate effect size; and a correlation coefficient of 0.50 or larger is considered to represent a strong or large effect size.

Effect size of correlation

2.3   Using R WebPower for Power Analysis for Correlation

The online power analysis is carried out using the R package WebPower on our Web server. The package can be directly used within R for power analysis for correlation. Specifically, the function wp.correlation is used. The function input is

```
wp.correlation(n = NULL, r = NULL, power = NULL, p = 0, rho0 = 0,
    alpha = 0.05, alternative = c("two.sided", "less", "greater
    "))
```

n: sample size
r: correlation or effect size
power: statistical power
p: number of variables to partial out
rho0: null correlation coefficient
alpha: significance level
alternative: alternative hypothesis

The R function can be used to conduct the same power analysis conducted by the online WebPower. For example, the R input and output for calculating power in Example 2.1.1 are given below.

Correlation Coefficient

Figure 2.1.5: Input and output for sample size planning for correlation in Example 2.1.3

Parameters (Help)

Sample size

Correlation   0.3

# of vars partialed out   0

Significance level   0.05

Power   0.8

H1   Two sided ⬍

Power curve   No power curve   ⬍

Note   Power for correlation

Calculate

Output

```
Power for correlation

      n    r alpha power
   83.95 0.3  0.05    0.8


URL: http://psychstat.org/correlation
```

```
> wp.correlation(n=50, r=0.3, alternative = "two.sided")

Power for correlation

    n   r alpha      power
   50 0.3  0.05 0.5728731

WebPower URL: http://psychstat.org/correlation
```

The R input and output for generating the power curve in Example 2.1.2 are given below.

Note that to generate the power plot, one first saves the power analysis results to example and then uses the plot function to obtain the plot.

```
> example = wp.correlation(n=seq(50,100,10), r=0.3, alternative =
    "two.sided")
> example

Power for correlation

    n   r alpha      power
   50 0.3  0.05 0.5728731
   60 0.3  0.05 0.6541956
   70 0.3  0.05 0.7230482
   80 0.3  0.05 0.7803111
```

Correlation Coefficient

Figure 2.1.6: Input and output for calculating minimum detectable effect size in Example 2.1.4

Parameters (Help)

Sample size   50

Correlation

# of vars partialed out   0

Significance level   0.05

Power   0.8

H1   Two sided ⬍

Power curve   No power curve ⬍

Note   Power for correlation

Calculate

Output

```
Power for correlation

    n      r alpha power
   50 0.3838  0.05   0.8


URL: http://psychstat.org/correlation




    90 0.3  0.05 0.8272250
   100 0.3  0.05 0.8651692


WebPower URL: http://psychstat.org/correlation


## to generate the power curve, simply use the plot function
> plot(example,type='b')
```

We could also use R to estimate both the sample size in Example 2.1.3 and the minimum detectable effect size in Example 2.1.4.

```
> wp.correlation(n=NULL,r=0.3, power=0.8, alternative = "two.
    sided")

Power for correlation

        n   r alpha power
  83.94932 0.3  0.05   0.8

WebPower URL: http://psychstat.org/correlation

> wp.correlation(n=50, r=NULL, power=0.8, alternative = "two.
    sided")
```

Correlation Coefficient

Parameters (Help)

Sample size	50
Correlation	0.24
# of vars partialed out	1
Significance level	0.05
Power	
H1	Two sided ⬍
Power curve	No power curve ⬍
Note	Power for correlation

Calculate

Output

```
Power for correlation

    n     r alpha  power
   50 0.24  0.05 0.3889


URL: http://psychstat.org/correlation
```

```
Power for correlation

    n         r alpha power
   50 0.3838075  0.05   0.8

WebPower URL: http://psychstat.org/correlation
```

Finally, to conduct power analysis for partial correlation in Example 2.1.5, the input and output are shown below.

```
> wp.correlation(n=50, r=0.3, p=1, alternative = "two.sided")

Power for correlation

    n   r alpha     power
   50 0.3  0.05 0.5640394

WebPower URL: http://psychstat.org/correlation
```

2.4   Technical Details

The power calculation for correlation is conducted based on Fisher's z transformation of the Pearson correlation coefficient. Fisher's z transformation of the sample correlation $r_{Y,X|W}$ is defined as

$$z = \frac{1}{2} \log \left(\frac{1 + r_{Y,X|W}}{1 - r_{Y,X|W}} \right).$$

After transformation, the statistic z approximately follows a normal distribution $N(\mu, \sigma^2)$, where

$$\mu = \frac{1}{2} \log \left(\frac{1 + \rho_{Y,X|W}}{1 - \rho_{Y,X|W}} \right) + \frac{\rho_{Y,X|W}}{2(N - 1 - p*)}$$

and

$$\sigma^2 = \frac{1}{N - 3 - p*},$$

where N is the sample size, $p*$ is the number of variables partialed out, and $\rho_{Y,X|W}$ is the population partial correlation coefficient between Y and X adjusting for the set of zero or more variables W.

The R function `wp.correlation` is based on the test statistic

$$z* = (N - 3 - p*)^{\frac{1}{2}} \left[z - \frac{1}{2} \log \left(\frac{1 + \rho_0}{1 - \rho_0} \right) - \frac{\rho_0}{2(N - 1 - p*)} \right],$$

where ρ_0 is the null partial correlation coefficient. $z*$ follows a normal distribution $N(\delta, v)$ with δ and v defined as

$$\delta = (N - 3 - p*)^{\frac{1}{2}} \left\{ \frac{1}{2} \log(\frac{1 + \rho_{Y,X|W}}{1 - \rho_{Y,X|W}}) + \frac{\rho_{Y,X|W}}{2(N - 1 - p*)} \left[1 + \frac{5 + \rho_{Y,X|W}^2}{4(N - 1 - p*)} + \frac{11 + 2\rho_{Y,X|W}^2 + 3\rho_{Y,X|W}^4}{8(N - 1 - p*)^2} \right] \right.$$
$$\left. - \frac{1}{2} log(\frac{1 + \rho_0}{1 - \rho_0}) - \frac{\rho_0}{2(N - 1 - p*)} \right\}$$

$$v = \frac{N - 3 - p*}{N - 1 - p*} \left[1 + \frac{4 - \rho_{Y,X|W}^2}{2(N - 1 - p*)} + \frac{22 - 6\rho_{Y,X|W}^2 - 3\rho_{Y,X|W}^4}{6(N - 1 - p*)^2} \right].$$

For more details, see Kendall et al. (1994). Based on the normal distribution, the statistical power is computed as

$$power = \begin{cases} \Phi\left[(\delta - z_{1-\alpha})/\sqrt{v} \right] & \text{upper one-sided} \\ \Phi\left[(-\delta - z_{1-\alpha})/\sqrt{v} \right] & \text{lower one-sided} \\ \Phi\left[(\delta - z_{1-\alpha/2})/\sqrt{v} \right] + \Phi\left[(-\delta - z_{1-\alpha/2})/\sqrt{v} \right] & \text{two-sided} \end{cases},$$

where Φ is the normal distribution function and z_α is the 100αth percentile of the standard normal distribution.

2.5   *Exercises*

1. A researcher plans to investigate the relationship between quality of life and family income. Based on his prior knowledge, these two variables are correlated at 0.2. What would be the required sample size if he wants to get a power of 0.8 with the alpha level set at 0.05?

2. Using the same information in Exercise 1, generate a power curve with the total sample size ranging from 100 to 300 with an interval of 10. From the power curve, approximately how large a sample size is needed to get a power 0.9?

3. Using the same information in Exercise 1, what would be the required sample size when the alpha level is set at 0.1 and 0.01, respectively?

4. In addition to the relationship between quality of life and family income, the expenditure on social services may also be correlated with these two variables. After taking into account the effect of expenditure on social services, the partial correlation between quality of life and family income will be 0.14 based on the prior information. What would be the power if he still uses the sample size obtained in Exercise 1?

3 | *Statistical Power Analysis for Tests of Proportions*

Meghan K. Cain
Department of Psychology
University of Texas at San Antonio

Tests of proportions are a technique used to compare proportions of success or agreement, p, in one or two samples. "Success" can literally be successful completion of a test question, or the researcher can designate one of the possible outcomes as success and model the proportion of participants that choose or fall into that particular outcome. It is not necessary to model the proportion choosing the other option, as it must be $1 - p$. The one-sample test of proportion tests the null hypothesis that the sample is drawn from a population with the proportion of success under the null hypothesis, usually 0.5, and the two-sample test of proportions tests the null hypothesis that the two samples are drawn from populations with the same proportion of success. In this book, the Z-test is often used to evaluate whether the given difference in proportions is significantly different from the null, although other methods are available.

3.1 | *One-sample Proportion Test*

We first demonstrate how to conduct statistical power analysis for the one-sample proportion test. The primary software interface of WebPower for power analysis for a one-sample test of proportion is shown in Figure 3.1.1. Within the interface, a user can supply different parameter values and select different options for power analysis. Among the four parameters, *Sample size*, *Effect size*, *Significance level*, and *Power*, one and only one can be left blank.

URL: http://psychstat.org/prop

- The *Sample size* is the total number of participants in the sample. Multiple sample sizes can be provided in two ways. First, multiple sample sizes can be supplied and separated by white spaces. For example, "100 150 200" will calculate power for the three sample

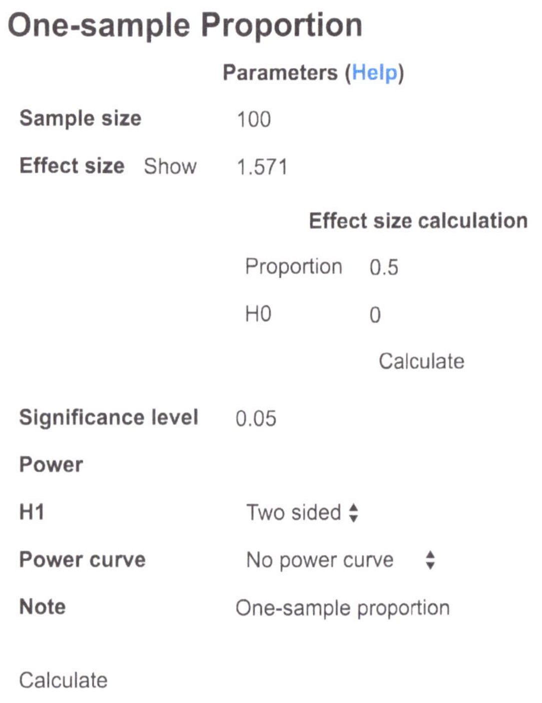

One-sample Proportion

Parameters (Help)

Sample size	100
Effect size Show	1.571

Effect size calculation

Proportion	0.5
H0	0
	Calculate

Significance level	0.05
Power	
H1	Two sided ⬍
Power curve	No power curve ⬍
Note	One-sample proportion

Calculate

Figure 3.1.1: Interface of power calculator for one-sample proportion

sizes 100, 150 and 200. A sequence of sample sizes can also be generated using the method *s:e:i* with *s* denoting the starting sample size, *e* denoting the ending sample size, and *i* as the interval. For example, "100:150:10" will calculate power for the sample sizes 100, 110, 120, 130, 140 and 150. By default, the sample size is 100. Alternatively, this field can be left blank and the required sample size can be calculated for a given power, effect size, and significance level.

- The *Effect size* is a measure of how different the proportion under evaluation is from the proportion under the null hypothesis. It can be input directly or can be calculated by clicking the "Show" button and then inputting the proportions as in Figure 3.1.1. Note that the proportion itself is not the effect size used in calculation. The expected proportion must be supplied in the "Proportion" box and the proportion under the null hypothesis must be supplied in the "H0" box. Once these two values have been supplied, the user can click "Calculate" and the effect size will be automatically calculated and filled in on the primary interface page. Cohen (1992) suggests that effect size values of 0.2, 0.5, and 0.8 represent "small", "medium", and "large" effect sizes, respectively, for tests of proportion. Multiple effect sizes can beprovided using the same method as sample sizes. Alternatively, this field can be left blank to calculate the minimum detectable effect size for a given sample size, significance level, and

Effect size *h* for proportion

power.

- The *Significance level*, also called the alpha level or nominal type I error rate, is specified as 0.05 by default as is a convention in many fields. This means that a significant result will be found only 5% of the time when the population proportion is equal to the proportion under the null hypothesis. Smaller values may be used with multiple comparisons to manage type I error rates across analyses appropriately.

- The *Power* is the desired power level that a user would like to obtain, usually 0.80. It can be left blank to have power calculated given sample size, effect size, and significance level. Power is the probability of obtaining a significant result when the population proportion of success is not equal to the proportion under the null hypothesis specified. The H1 can be specified as "Two-sided", "Less", or "Greater" than the proportion under the null hypothesis. The corresponding alternative hypotheses are:

$$
\begin{aligned}
\text{``Two sided''}: & \quad \pi \neq p_0 \\
\text{``Less''}: & \quad \pi < p_0 \\
\text{``Greater''}: & \quad \pi > p_0,
\end{aligned}
\tag{3.1.1}
$$

where π is the population proportion and p_0 is the proportion under the null hypothesis. Research hypotheses containing the phrase "at least as" usually refer to the "Greater" alternative hypothesis, and those containing the phrase "at most as" usually refer to the "Less" alternative hypothesis.

- Whether to generate a power curve can be specified in the drop-down menu next to the *Power curve* option. Power curves are useful when deciding on an appropriate sample size for a planned study or when examining power at feasible effect sizes. They either show corresponding powers for many sample sizes given an effect size, or for many effect sizes given a sample size.

- The *Note* option will title and save the current power calculation for future reference if the user has logged in to the website. This note must be less than 200 characters.

Once all fields have been appropriately filled in, pressing the "Calculate" button will create a table with the calculated results such as power, effect size, sample size, and if specified, a power curve will appear below the table.

3.1.1 Examples

The mayor of Powerville is debating passing a new law in his town. Suppose that 300 residents of this small town are surveyed and asked whether they would support the passing of this law. What is the power of the test if 168 residents are expected to report that they would support the passing of the law?

The input and output for calculating power for this study are given in Figure 3.1.2. The *Sample size* is specified as 300. The *Effect size* is calculated to be 0.120 by inputting the proportion, 168/300=0.56[1], and the proportion under the null hypothesis, 0.5, after clicking the "Show" button. The proportion under the null hypothesis is 0.5 because this would correspond to an equal number of citizens supporting and not supporting this law, as would be expected if there was no preference. The default *Significance level* of 0.05 is used to maintain the alpha level at 0.05. The *Power* is left blank because this is what the mayor is interested in calculating. He is just interested in whether there is a difference so he keeps the *H1* as two-sided. A simple note, "Powerville Bill", is added in the *Note* field in case he wants to refer back to this calculation again in the future. He clicks "Calculate" and sees that he has a power of 0.5472 in this test.

A student is wondering whether there are more republicans or democrats at his school. For the purpose of this study, no other options are provided. He is not sure how large of a difference there will be or in what direction, but he knows he will only have time to poll 100 students. What will his power be for this study at different effect sizes? Is it a study worth pursuing?

The input and output for calculating power for this study are given in Figure 3.1.3. The *Sample size* is specified as 100. The *Effect size* ranges from a small effect size of 0.2 to a large effect size of 0.8, since he is unsure how large it will be. The default *Significance level* of 0.05 is used to maintain his alpha level at 0.05. The *Power* is left blank since this is what he's interested in calculating. He is just interested in whether there is a difference, so he keeps the *H1* as two-sided. He selects "Show power curve" so that he can use the curve to help him decide whether the study is worthwhile. A simple note, "Repo vs. Demo" is added in the *Note* field in case he wants to reference this calculation again in the future. He clicks "Calculate" and sees from the power curve in Figure 3.1.4 that as long as the effect size is about 0.3, he should have enough power. Given the proportion under the null hypothesis 0.50 (equal amounts of republicans and democrats), this effect size corresponds to a population proportion of about 65%. This proportion seems feasible

Example 3.1.1: Calculate power given sample size and effect size

[1] In practice, since the population value is not known, it is often estimated using the sample value. But the estimated value might not be the same as the population value.

Example 3.1.2: Power curve with different effect sizes

One-sample Proportion

Figure 3.1.2: Input and output for calculating power for a one-sample proportion test in Example 3.1.1

Parameters (Help)

Sample size   300

Effect size Show   0.120

Effect size calculation

Proportion   0.56

H0   0.5

Calculate

Significance level   0.05

Power

H1   Two sided ⬍

Power curve   No power curve   ⬍

Note   Powerville Bill

Calculate

Output

```
Power for one-sample proportion test

      h   n alpha  power
   0.12 300  0.05 0.5472

URL: http://psychstat.org/prop
```

for his school, so he decides to go ahead with his study.

Example 3.1.3: Calculate sample size given power and effect size

A new drug needs to be tested before it can be put on the market. The proportion of users reporting improvement with the best drug already on the market is 60%, and drug-developers are interested in whether their drug performs significantly better than this. Pilot clinical trials have reported that 70% of users have shown improvement. How many participants must be in this clinical trial to have a power of 0.80 of finding a significant improvement over the drug already on the market?

The input and output for calculating power for this study are given in Figure 3.1.5. The *Sample size* is left blank because the researchers are interested in calculating how many participants are necessary for this study. The *Effect size* is calculated as follows. The proportion of effectiveness for the new drug is specified as 0.70 to match early clinical trial estimates. The proportion under the null hypothesis is specified

One-sample Proportion

Parameters (Help)

Sample size   100

Effect size Show   0.2:0.8:0.1

Significance level   0.05

Power

H1   Two sided ↕

Power curve   Show power curve ↕

Note   Repo vs. Demo

Calculate

Output

```
Power for one-sample proportion test

    h   n alpha  power
  0.2 100  0.05 0.5160
  0.3 100  0.05 0.8508
  0.4 100  0.05 0.9793
  0.5 100  0.05 0.9988
  0.6 100  0.05 1.0000
  0.7 100  0.05 1.0000
  0.8 100  0.05 1.0000


URL: http://psychstat.org/prop
```

Figure 3.1.3: Input and output for calculating power curve for a one-sample proportion test in Example 3.1.2

as 0.60 to correspond to which we are interested in comparing to this sample, the current best drug. After pressing the "Calculate" button, we see that this corresponds to an effect size of 0.210. The default *Significance level* of 0.05 is used to maintain his alpha level at 0.05. The *Power* is specified as 0.80 to correspond to the desired power of this study. The tail is specified as "Greater" because we only care whether the new drug is better, not worse. A simple note, "New Drug" is added in the *Note* field. After pressing "Calculate" we see that we will need 141 participants (rounded to an integer) to have adequate power for this study.

3.1.2   *Using R package WebPower*

The online power calculator is based on the R function `wp.prop` in the WebPower package. The function input is

Figure 3.1.4: Power curve for a one-sample proportion test in Example 3.1.2

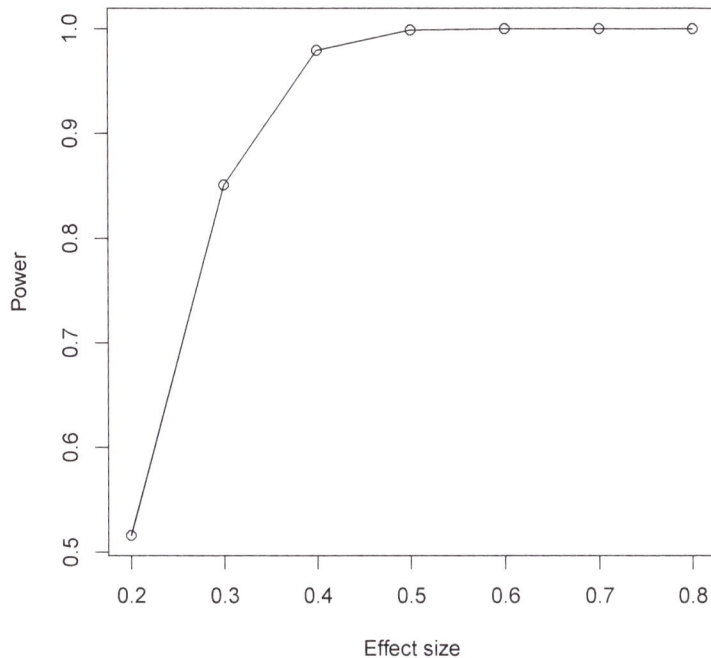

```
wp.prop(h = NULL, n1 = NULL, n2 = NULL, alpha = 0.05, power =
    NULL, type = c('1p', '2p', '2p2n'), alternative = c("two.
    sided", "less","greater"))
```

For one-sample proportion analysis, one can supply the sample size using n1. The default type of analysis is the one-sample proportion test. The R input and output below show how to conduct the power analysis for the examples discussed in the previous section.

```
> ## Calculate power
> wp.prop(h=.120, n1=300)

Power for one-sample proportion test

      h   n sig.level   power
   0.12 300      0.05 0.54719


>
> ## Generate a power curve
> res <- wp.prop(h=seq(0.2,0.8,0.1), n1=100)
> res

Power for one-sample proportion test

      h   n sig.level      power
```

h: effect size
n1: sample size for group 1
n2: sample size for group 2
alpha: significance level
power: statistical power
type: type of analysis. 1p, one sample; 2p, two sample with equal sample size; 2p2n, two sample with unequal sample size
alternative: alternative hypothesis

One-sample Proportion

Figure 3.1.5: Input and output for calculating sample size for a one-sample proportion test in Example 3.1.3

Parameters (Help)

Sample size

Effect size   Show   0.210

Effect size calculation

Proportion   0.7

H0   0.6

Calculate

Significance level   0.05

Power   0.8

H1   Greater   ↕

Power curve   No power curve   ↕

Note   New Drug

Calculate

Output

```
Power for one-sample proportion test

     h      n alpha power
  0.21 140.2  0.05   0.8


URL: http://psychstat.org/prop
```

```
   0.2 100     0.05 0.5160053
   0.3 100     0.05 0.8508388
   0.4 100     0.05 0.9793266
   0.5 100     0.05 0.9988173
   0.6 100     0.05 0.9999733
   0.7 100     0.05 0.9999998
   0.8 100     0.05 1.0000000

> plot(res, xvar='h', yvar='power')
>
> ## Estimate the sample size
> wp.prop(h=.210, n1=NULL, power=0.8, alternative='greater')

Power for one-sample proportion test

     h       n sig.level power
  0.21 140.194      0.05   0.8
```

3.1.3   Note on choosing a proportion under the null hypothesis and deciding between a one- or two-tailed test

The first step in an analysis of one proportion is to figure out the proportion under the null hypothesis of the population, p_0. For example, if two choices were presented to participants and under the null we would expect each choice to be just as likely, the corresponding proportion under the null hypothesis would be 0.5. If there were no hypothesis about whether one answer would be more likely than the other, a two-tailed test of proportion would be implemented. Otherwise, an upper tail can be used to test whether the proportion of success is higher than 0.5, and a lower tailed test can be used to test whether the proportion of success is lower than 0.5.

The proportion under the null hypothesis must not always be 0.5. For example, if the research question is whether at least 75% of a population can answer a certain question correctly, the corresponding statistical test would be comparing the proportion of success to $p_0 = 0.75$ and testing only the upper tail. Using different proportions under the null hypothesis with one- or two-tailed tests correspond to different research questions, thus care is advised when making these decisions.

3.1.4   Technical details

Under a two-tailed test, the null hypothesis in a one-sample test of proportion is that the population proportion of success, π, is equal to some specified proportion under the null hypothesis, p_0 (Eq. 3.1.2), and the alternative hypothesis is that they are not equal (Eq. 3.1.3).

$$H_0: \quad \pi = \quad p_0 \qquad (3.1.2)$$
$$H_a: \quad \pi \neq \quad p_0 \qquad (3.1.3)$$

Alternative hypotheses under one-tailed tests appear in Equation 3.1.1.

The power calculation is based on the arcsine transformation of the proportion (see Cohen, 1988; p548). Specifically, for a given proportion, p, the transformation $\phi(p)$ is

$$\phi(p) = 2 \times \arcsin\left(\sqrt{p}\right).$$

Given the proportion under the null hypothesis, p_0, the effect size h is defined by the difference after transformation,

$$h = \phi(p) - \phi(p_0) \qquad (3.1.4)$$
$$= 2 \times \arcsin\left(\sqrt{p}\right) - 2 \times \arcsin\left(\sqrt{p_0}\right).$$

The power is then calculated by

$$Power = \begin{cases} \Phi\left(z_{\alpha/2} - h\sqrt{n}\right) + 1 - \Phi\left(z_{1-\alpha/2} - h\sqrt{n}\right) & \text{two sided} \\ 1 - \Phi\left(z_{1-\alpha} - h\sqrt{n}\right) & \text{less} \\ \Phi\left(z_{\alpha} - h\sqrt{n}\right) & \text{greater} \end{cases},$$

(3.1.5)

where n is the sample size, Φ is the standard normal cumulative distribution function, and z_{α} is the critical value from the standard normal distribution. The higher the effect size, the lower Φ will be, and higher the power will be.

3.2 Two-sample Proportion Test with Equal Sample Sizes

The primary software interface for power analysis for a two-sample test of proportions with equal sample sizes is shown in Figure 3.2.1. As with the one-sample proportion test, among the four parameters, *Sample size*, *Effect size*, *Significance level*, and *Power*, one and only one can be left blank.

URL: http://psychstat.org/prop2p/

Two-sample Proportion

Parameters (Help)

Sample size	100
Effect size Show	0.120

Effect size calculation

Proportion 1	0.56
Proportion 2	0.5

Calculate

Significance level	0.05
Power	
H1	Two sided ↕
Power curve	No power curve ↕
Note	Two-sample proportion

Calculate

Figure 3.2.1: Interface of power calculator for two-sample proportions, equal sample sizes

- The *Sample size* is the number of participants in *each* sample. If using this calculator, the sample sizes of each sample should be the same. If they are not, please see Section 3.3 for the two-sample proportion test with unequal sample sizes. Multiple sample sizes can be provided as in the one-sample method. The sample size field can be left blank and the required sample size can be calculated for a given power, effect size, and significance level.

- The *Effect size* is a measure of the difference between two proportions. It can either be inputted directly or can be calculated by clicking the "Show" button and then inputting the corresponding proportions. The proportion of population 1, p_1, must be supplied in the "Proportion 1" box and the proportion of population 2, p_2, must be supplied in the "Proportion 2" box. If using a two-sided test, it is arbitrary to decide which is Proportion 1 and which is Proportion 2. However, if using one-sided test these labels must be assigned properly to align with the research hypothesis. Once these boxes have been filled out the user can click "Calculate" and the effect size will be automatically calculated and filled in on the primary interface page. Cohen suggests that effect size values of 0.2, 0.5, and 0.8 represent "small", "medium", and "large" effect sizes, respectively, for tests of proportion (Cohen, 1992). Multiple effect sizes can be provided using the same method as with sample sizes. Alternatively, this field can be left blank to calculate the minimum detectable effect size for a given sample size, significance level, and power.

- The *Significance level*, otherwise called the alpha level or nominal type I error rate, is specified as 0.05 by default as is a convention in many fields.

- The *Power* is the desired power level the user would like to obtain, usually 0.80, or can be left blank to have power calculated given sample size, effect size, and significance level.

- The H1 can be specified as "Two-sided", "Less", or "Greater". The corresponding alternative hypotheses are:

$$\begin{aligned} \text{"Two sided"}: \quad & \pi_1 = \pi_2 \\ \text{"Less"}: \quad & \pi_1 < \pi_2 \\ \text{"Greater"}: \quad & \pi_1 > \pi_2, \end{aligned} \tag{3.2.1}$$

where π_1 is the population proportion 1, and π_2 is the population proportion 2.

- Whether to generate a power curve can be specified in the drop-down menu next to the *Power curve* option. Power curves are useful when deciding on an appropriate sample size for a planned study or when examining power at feasible effect sizes. They either show corresponding powers for many sample sizes given an effect size, or for many effect sizes given a sample size.

- The *Note* option will title and save the current power calculation for future reference if the user has logged in to the website. This note must be less than 200 characters.

Once all fields have been appropriately filled in, pressing the "Calculate" button will create a table of results and, if specified, a power curve will appear below the table.

A public health researcher is curious to know how adolescent obesity rates in California compare to that in New York. He believes that if he surveys 100 students from various high schools in California and 100 students from high schools in New York, 25 students in California will be reported to be obese and 35 students in New York will be reported to be obese. What is the power of such a study?

The input and output for calculating power for this study are given in Figure 3.2.2. *Sample size* is 100 because the researcher will survey 100 students from *each* region. The *Effect size* is calculated using the obesity rates in California and New York. Note that since this is a two-tailed test, it does not matter which is *Proportion 1* and which is *Proportion 2*. The default *Significance level* of 0.05 is used to maintain the alpha level at 0.05. The *Power* is left blank since the researcher is interested in calculating the power for his study. The option *H1* is specified as "Two-sided" because prior to administering the survey, the researcher is simply interested in whether the two obesity rates differ, not in a specific rate being higher or lower. A simple note, "Obesity" is added in the *Note* field in case he wants to refer to this calculation again in the future. He clicks the *Calculate* button and sees that he has a power of about .3406 in this study.

The public health researcher thus found that he needs to collect more data on obesity rates if he wants to have enough power in his study. Using the same effect size he found in Example 3.2.1, he wants to know what the power would be at sample sizes up to 500 in increments of 100 to help him decide how many more students to survey.

The input and output for calculating power for this study are given in Figure 3.2.3. *Sample size* is "100:500:100" because the researcher is interested in calculating power for the sample sizes 100, 200, 300, 400, and 500. The *Effect size* is calculated using the obesity rates in California and New York that he found in his first study. The default *Significance level* of 0.05 is used to maintain the alpha level at 0.05. The *Power* is left blank since the researcher is interested in calculating the power for his study. The option *H1* is specified as "Two-sided" because

Example 3.2.1: Calculate power given two proportions with equal sample sizes

Example 3.2.2: Power curve with multiple sample sizes

Two-sample Proportion

Parameters (Help)

Sample size   100

Effect size Show   -0.219

Effect size calculation

Proportion 1   0.25

Proportion 2   0.35

Calculate

Significance level   0.05

Power

H1   Two sided ⬍

Power curve   No power curve ⬍

Note   Obesity

Calculate

Output

```
Power for two-sample proportion (equal n)

      h    n alpha  power
  0.219 100  0.05 0.3406


NOTE: Sample sizes for EACH group
URL: http://psychstat.org/prop2p
```

Figure 3.2.2: Input and output for calculating power for a balanced two-sample proportion test in Example 3.2.1

the researcher is still only interested in whether the two obesity rates differ. He specified that he wants to "Show power curve" so that he can visually inspect the relationship between sample size and power in his study. A simple note, "Obesity sample sizes" is added in the *Note* field. He clicks the *Calculate* button and sees that the table of power values in Figure 3.2.3 and the power curve in Figure 3.2.4. He wants to have a power of at least .80, for which he sees he will need a sample size between 300 and 400. He decides to collect data from 400 students to be assured that he will have enough power for his study.

Two-sample Proportion

Parameters (Help)

Sample size	100:500:100
Effect size Show	-0.219
Significance level	0.05
Power	
H1	Two sided ⬍
Power curve	Show power curve ⬍
Note	Obesity sample sizes

Calculate

Output

```
Power for two-sample proportion (equal n)

      h   n alpha  power
  0.219 100  0.05 0.3406
  0.219 200  0.05 0.5910
  0.219 300  0.05 0.7649
  0.219 400  0.05 0.8723
  0.219 500  0.05 0.9335


NOTE: Sample sizes for EACH group
URL: http://psychstat.org/prop2p
```

Figure 3.2.3: Input and output for obtaining a power curve for a two-sample proportion test in Example 3.2.2

3.2.1 *Use R package for two-sample proportion test with equal sample sizes*

The analysis in Examples 3.2.1 and 3.2.2 can also be conducted in R using the wp.prop function as shown below. Note that the type of analysis is set to be "2p".

```
> wp.prop(h=.219, n1=100, type='2p')

Power for two-sample proportion test (equal sample size)

      h   n alpha     power
  0.219 100  0.05 0.3406149

> res <- wp.prop(h=.219, n1=seq(100,500,100), type='2p')
> res

Power for two-sample proportion test (equal sample size)
```

Figure 3.2.4: Power curve for a two-sample proportion test in Example 3.2.2

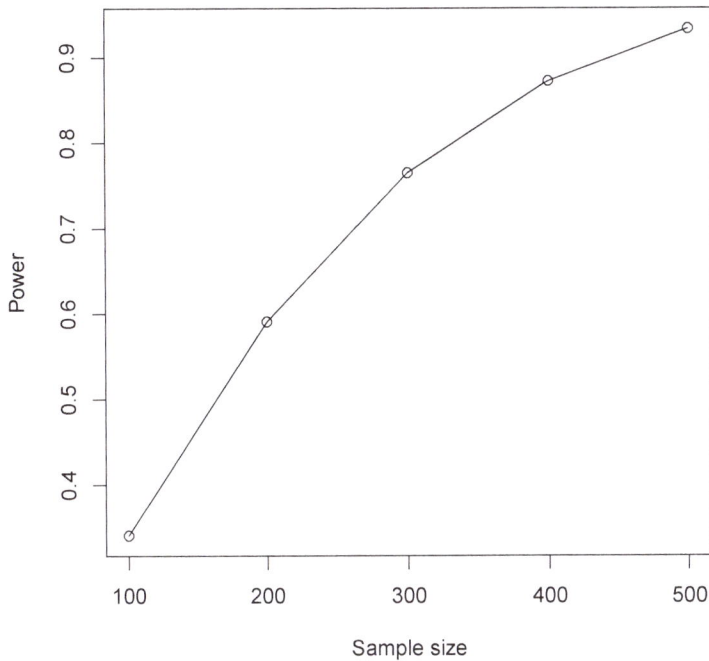

```
     h   n alpha     power
 0.219 100  0.05 0.3406149
 0.219 200  0.05 0.5909847
 0.219 300  0.05 0.7649243
 0.219 400  0.05 0.8722653
 0.219 500  0.05 0.9335457
```

3.2.2   Technical details

Under a two-tailed test, the null hypothesis in a two-sample test of proportions is that the population 1 proportion of success, π_1, is equal to the population 2 proportion of success, π_2 (Eq. 3.2.2), and the alternative hypothesis is that they are not equal (Eq. 3.2.3).

$$H_0 : \quad \pi_1 = \quad \pi_2 \tag{3.2.2}$$

$$H_a : \quad \pi_1 \neq \quad \pi_2 \tag{3.2.3}$$

Alternative hypotheses under one-tailed tests appear in Equation 3.2.1.

The effect size is calculated in the same way as in the one-sample test using the arcsin transformation, except that the proportion from one population, p_1, is compared to another, p_2, instead of to a proportion under the null hypothesis. Therefore, the effect size, h, can be calculated as

$$h = 2 \times \arcsin\left(\sqrt{p_1}\right) - 2 \times \arcsin\left(\sqrt{p_2}\right). \qquad (3.2.4)$$

Choosing which proportion is first or second will only affect power calculations in the case of "Greater" or "Less" than tails, otherwise they are arbitrary. The power calculation is performed in the same way as in Equation 3.1.5:

$$Power = \begin{cases} \Phi\left(z_{\alpha/2} - h\sqrt{n}\right) + 1 - \Phi\left(z_{1-\alpha/2} - h\sqrt{n}\right) & \text{two sided} \\ 1 - \Phi\left(z_{1-\alpha} - h\sqrt{n}\right) & \text{less} \\ \Phi\left(z_{\alpha} - h\sqrt{n}\right) & \text{greater} \end{cases},$$

(3.1.5; Reproduced)

where n is the sample size of *each* sample, Φ is the standard normal cumulative distribution function, and z_{α} is the critical value from the standard normal distribution.

3.3   *Two-sample Proportion Test with Unequal Sample Sizes*

The primary software interface for power analysis for a two-sample test of proportion with unequal sample sizes is shown in Figure 3.3.1. Within the interface, a user can supply different parameter values and select different options for power analysis. Among the five parameters, *Sample size 1*, *Sample size 2*, *Effect size*, *Significance level*, and *Power*, one and only one can be left blank.

http://psychstat.org/prop2p2n

- *Sample size 1* is the number of participants in the first sample, and *Sample size 2* is the number of participants in the second sample. By default, both sample sizes are set to 100. Either *Sample size 1* or *Sample size 2* can be left blank and the required sample size for a group can be calculated.

- The *Effect size* is a measure of how different proportion 1 is from proportion 2. It can either be inputted directly or can be calculated by clicking the "Show" button and then inputting the proportions. The proportion 1, p_1, must be supplied in the "Proportion 1" box and the proportion, p_2, must be supplied in the "Proportion 2" box. Once these boxes have been filled out, the user can click "Calculate" and the effect size will be automatically calculated and filled in. Cohen suggests that effect size h values of 0.2, 0.5, and 0.8 represent "small", "medium", and "large" effect sizes, respectively, for tests of proportion (Cohen, 1992). Multiple effect sizes can be inputted in the same way that multiple sample sizes can. Alternatively, this field can be left blank to calculate the least required effect size for a given sample size, significance level, and power.

Two-sample unbalanced proportion

Parameters (Help)

Sample size 1	100
Sample size 2	100
Effect size Show	0.5

Effect size calculation

Proportion 1   0.6

Proportion 2   0.5

Calculate

Significance level	0.05
Power	
H1	Two sided ⬍
Power curve	No power curve ⬍
Note	Two-sample unbalance

Calculate

Figure 3.3.1: Interface of power calculator for two-sample proportions, unequal sample sizes

- The *Significance level*, otherwise called the alpha level or nominal type I error rate, is specified as 0.05 by default as is a convention in Psychology and other fields.

- The *Power* is the desired power level the user would like to obtain, usually 0.80, or can be left blank to have power calculated given sample size, effect size, and significance level. Power is the probability of obtaining a significant result when the population proportion of success in Sample 1 is not equal to that of Sample 2. The H1 can be specified as "Two-sided", "Less", or "Greater". The corresponding alternative hypotheses are:

$$\text{"Two sided"} : \pi_1 = \pi_2 \qquad (3.2.1, \text{ Reproduced})$$
$$\text{"Less"} : \pi_1 < \pi_2$$
$$\text{"Greater"} : \pi_1 > \pi_2,$$

where π_1 is the population proportion of Sample 1, and π_2 is the population proportion of Sample 2.

- Whether or not to generate a power curve can be specified in the drop-down menu next to the *Power curve* option. Power curves are useful when deciding on an appropriate sample size for a planned study or when examining power at feasible effect sizes. They either show corresponding powers for many sample sizes given an effect size, or for many effect sizes given a sample size.

- The *Note* option will title and save the current power calculation for future reference if the user has logged in to the website. This note must be less than 200 characters.

Once all fields have been appropriately filled in, clicking "Calculate" will create a table and, if specified, a power curve will appear below the table.

3.3.1 Examples

A teacher would like to know whether her honors class has learned the material significantly better than another one of her classes. She has designed a test and would like to compare how many pass in one class compared to the other. She estimates that about 90% will pass in the honors class and that 70% will pass in the other class. Unfortunately, the class sizes are not equal. She has 35 students in the honors class, and 50 students in the other class. Will she have enough power for her study?

The input and output for calculating power for this study are given in Figure 3.3.2. *Sample size* 1 is 35 because she has 35 students in her honors class, and *Sample size 2* is 50 because she has 50 students in her other class. The *Effect size* is calculated using the expected pass rates for each of the classes, which comes out to 0.52. Notice that *Proportion 1* must match the class that has *Sample size 1*, and *Proportion 2* must match the class with *Sample size 2*. The default *Significance level* of 0.05 is used to maintain the alpha level at 0.05. The *Power* is left blank since she is interested in calculating the power for her study. The option *H1* is specified as "Greater" because she would like to test whether her honors class will have a greater pass rate than her other class. She presses "Calculate" and sees that she will have a power of about 0.7626 for her study.

Example 3.3.1: Calculate power given two proportions with unequal sample sizes

3.3.2 R code for power calculation

The R input and output for Example 3.3.1 is given below.

```
> wp.prop(h=.52, n1=35, n2=50, alternative="greater",
        type="2p2n")


Power for two-sample proportion test (unequal sample size)
```

Two-sample unbalanced proportion

Parameters (Help)

Sample size 1 35

Sample size 2 50

Effect size Show 0.52

Effect size calculation

Proportion 1 0.9

Proportion 2 0.7

Calculate

Significance level 0.05

Power

H1 Greater $\updownarrow$

Power curve No power curve $\updownarrow$

Note Pass Rates

Calculate

Output

```
Power for two-sample proportion (unequal n)

     h n1 n2 alpha  power
  0.52 35 50  0.05 0.7626


NOTE: Sample size for each group
URL: http://psychstat.org/prop2p2n
```

```
     h n1 n2 alpha     power
  0.52 35 50  0.05 0.7625743
```

3.3.3 *Technical details*

Under a two-tailed test, the null hypothesis in a two-sample test of proportions is that the population proportion of success of sample 1, π_1, is equal to the population proportion of success of sample 2, π_2 (Eq. 3.2.2), and the alternative hypothesis is that they are not equal (Eq. 3.2.3). Alternative hypotheses under one-tailed tests appear in Equation 3.2.1. The effect size is calculated the same way as in the two-sample equal sample size case,

$$h = 2 \times \arcsin\left(\sqrt{p_1}\right) - 2 \times \arcsin\left(\sqrt{p_2}\right). \qquad \text{(3.2.4; Reproduced)}$$

Because there are two sample sizes, the sample size of the first sample n_1 and the sample size of the second sample n_2 are used to calculate the harmonic sample size. The harmonic sample size can be calculated as

$$\tilde{n} = \frac{2n_1 n_2}{n_1 + n_2}. \qquad (3.3.1)$$

The power is then calculated as

$$Power = \begin{cases} \Phi\left(z_{\alpha/2} - h\sqrt{\tilde{n}}\right) + 1 - \Phi\left(z_{1-\alpha/2} - h\sqrt{\tilde{n}}\right) & \text{two sided} \\ 1 - \Phi\left(z_{1-\alpha} - h\sqrt{\tilde{n}}\right) & \text{less} \\ \Phi\left(z_{\alpha} - h\sqrt{\tilde{n}}\right) & \text{greater} \end{cases},$$

$$(3.3.2)$$

where $\tilde{n}$ is the harmonic sample size, Φ is the standard normal cumulative distribution function, and z is the critical value from the standard normal distribution.

3.4   Exercises

1. A researcher plans to compare the proportions of female students in two colleges, UCLA and USC. To help decide the sample size, he has found the following data from corresponding school newspapers that he can use to calculate the expected population effect size. What would be the sample size needed for each school to get a power of 0.8 at the alpha level 0.05?

	UCLA	USC
Number of female students	26	42
Number of male students	35	39

2. Using the same information in Exercise 1, what would be the required sample sizes when the alpha level is set at 0.1? at 0.01?

3. If a researcher can collect two times more data from UCLA than from USC, to get a power 0.8, how many participants are needed in each sample?

4 Statistical Power Analysis for t-tests

Han Du
Department of Psychology
University of California, Los Angeles

A *t*-test is a statistical hypothesis test in which the test statistic follows a Student's *t*-distribution if the null hypothesis is true and follows a non-central *t* distribution if the alternative hypothesis is true. The *t*-test can assess whether a population mean, the difference between two independent population means, or the difference between means of matched pairs (dependent population means) equals a specific value.

4.1 How to Conduct Power Analysis for *t* tests

Power analysis for *t*-tests can be conducted using the online software WebPower with the interfaces shown in Figure 4.1.1 for one-sample or two-sample paired/balanced *t*-test and in Figure 4.1.2 for two-sample unbalanced *t*-test . There are four essential parameters, *Sample size*, *Effect size*, *Significance level*, and *Power*, in both balanced and unbalanced *t*-tests power analysis interfaces. With any three of them known, the fourth one can be calculated. Therefore, in addition to power calculation, sample size planning can also be conducted.

http://psychstat.org/ttest

http://psychstat.org/ttest2n

- *Sample size* specifies the number of observations per group. Multiple sample sizes can be provided in two ways. First, multiple sample sizes can be supplied and separated by white spaces (e.g., 100 150 200). Then power will be calculated for all these sample sizes. Second, a sequence of sample sizes can be generated using the method `s:e:i` with s denoting the starting sample size, e the ending sample size, and i the interval. For example, 100:150:10 will generate a sequence 100 110 120 130 140 150. For the one-sample *t*-test, paired *t*-test and balanced *t*-test, only one sample size is required (Figure 4.1.1), and for the unbalanced two-sample *t*-test, the sample sizes for both

One-sample, paired, two-sample balanced t test

Figure 4.1.1: Software interface for one-sample or two-sample paired/balanced *t* test

Parameters (Help)

Sample size	100
Effect size (Calculator)	0.1
Significance level	0.05
Power	
Type of test	Two sample ⬍
H1	Two sided ⬍
Power curve	No power curve ⬍
Note	t-test

Calculate

groups should be specified (Figure 4.1.2). By default, the sample size is 100 as shown in Figure 4.1.1.

- *Effect size* specifies the population difference. In practice, the population effect size can be hypothesized at the sample effect size, which should be used with caution. Multiple effect sizes or a sequence of effect sizes can be supplied using the same method as for sample size. By default, the value is 0.1. Without an effect size in hand, one may need to calculate the effect size first by clicking the link "Calculator".

- *Significance level* (Type I error rate) tells the alpha level for power calculation with the default value 0.05.

- *Power* specifies the required statistical power or can be left blank for calculation.

- One-sample *t*-test, paired *t*-test, and balanced two-sample *t*-test can be chosen through *Type of test* in the software interface of "Power of t-test" (Figure 4.1.1). Unbalanced two-sample *t*-test is done in a separate interface as shown in Figure 4.1.2.

- One can specify the alternative hypothesis through *H1*: "Two-sided" (default), "Greater" or "Less".

- A power curve can be plotted if multiple sample sizes or effect sizes are provided by choosing "Show power curve" in the drop-list of *Power curve*.

Unbalanced two-sample t test

Figure 4.1.2: Software interface for unbalanced two-sample *t* test

Parameters (Help)	
Sample size of group 1	100
Sample size of group 2	100
Effect size (Calculator)	0.1
Significance level	0.05
Power	
H1	Two sided ⬍
Power curve	No power curve ⬍
Note	Unbalanced two-sample

Calculate

Figure 4.1.2: Software interface for unbalanced two-sample *t* test

After providing the required information, a user can get the power analysis output by clicking the **Calculate** button.

4.1.1   *Examples*

A researcher is interested in whether the population mean (μ) of an experimental group is different from 0. If the effect size is known as 0.2 and there are 150 participants in the experimental group, what is the power to find the significant difference between μ and 0?

To determine a power, use the interface in Figure 4.1.1. Then, specify *Type of test* to be "One sample" to conduct the analysis. The input and output for calculating power for this study are given in Figure 4.1.3. In the field of *Sample size*, input 150, and in the field of *Effect size*, input 0.2, the known effect size. The default significance level 0.05 is used. Since we need to calculate power, the field for *Power* is left blank. By clicking the "**Calculate**" button, the statistical power is given in the output immediately. For this study, the power is 0.6822.

Patients' propensity scores are recorded before and after undergoing a new therapy. If the effect size is known as -0.4 and there are 40 patients, what is the power to find a significant decrease in propensity scores after the psychotherapy?

For this analysis, we need to use a paired t-test. Set *Type of test* to be "Paired" and choose *H1* to "Less". The input and output for calculating power for this study are given in Figure 4.1.4. In the field of *Sample size*, input. 40, and in the field of *Effect size*, input -0.4, the known effect size. The default significance level 0.05 is used. Since we need

Example 4.1.1: Calculate power given sample size and effect size in one-sample t-test

Example 4.1.2: Calculate power given sample size and effect size in paired t-test

One-sample, paired, two-sample balanced t test

Parameters (Help)

Sample size	150
Effect size (Calculator)	0.2
Significance level	0.05
Power	
Type of test	One sample ⇕
H1	Two sided ⇕
Power curve	No power curve ⇕
Note	t-test

Calculate

Output

```
One-sample t-test

     n   d alpha  power
   150 0.2  0.05 0.6822


URL: http://psychstat.org/ttest
```

Figure 4.1.3: Input and output for calculating power for one-sample *t*-test in Example 4.1.1

to calculate power, the field for *Power* is left blank. By clicking the "**Calculate**" button, the statistical power is given in the output, 0.7997 for this study.

A study is designed with 70 participants completing a task with disturbance and 70 participants without disturbance. If the effect size is known as 0.3, what is the power to find that the participants without disturbance perform significantly better than the ones with disturbance?

The two-sample t-test is used here by specifying *Type of test* to be "Two sample" and choosing *H1* to be "Greater". The input and output for calculating power for this study are given in Figure 4.1.5. In the field of *Sample size*, input 70, and in the field of *Effect size*, input 0.3, the known effect size. The default significance level 0.05 is used. Since we need to calculate power, the field for *Power* is left blank. By clicking the "**Calculate**" button, the statistical power is given in the output and it is 0.5483 for this study.

There are two groups of students. Students in group A use learning

Example 4.1.3: Calculate power given sample size and effect size in two-sample t-test

Example 4.1.4: Calculate power given sample size and effect size in unbalanced two-sample t-test

One-sample, paired, two-sample balanced t test

Parameters (Help)

Sample size 40

Effect size (Calculator) -0.4

Significance level 0.05

Power

Type of test Paired ⬍

H1 Less ⬍

Power curve No power curve ⬍

Note t-test

Calculate

Output

```
Paired t-test

    n    d alpha  power
   40 -0.4  0.05 0.7997


NOTE: n is number of *pairs*
URL: http://psychstat.org/ttest
```

Figure 4.1.4: Input and output for calculating power for paired *t*-test in Example 4.1.2

strategy A and students in group B use learning strategy B. If the effect size is known as 0.356 and one researcher plans to recruit 30 students for group A and 40 students for group B, what is the power to find a significant difference between the two groups in the study?

We need to use an unbalanced two-sample *t*-test in Figure 4.1.2 to conduct the analysis. The input and output for calculating power for this study are given in Figure 4.1.6. In the field of *Sample size of group 1*, input 30, and in the field of *Sample size of group* 2, input 40, and in the field of *Effect size*, input 0.356, the known effect size. The default significance level 0.05 is used. By clicking the "**Calculate**" button, we obtain a power 0.3065.

If the researcher does not know the effect size in Example 4.1.4 but expects the effect size to be in the range of 0.2 to 0.8 (from small effect size to large effect size according to Cohen, 1988), then the researcher can obtain a power curve with each potential value of the effect size.

Figure 4.1.7 shows the input and output for a study with 30 par-

Example 4.1.5: Power curve given sample size and effect size in unbalanced two-sample t-test

One-sample, paired, two-sample balanced t test

Figure 4.1.5: Input and output for calculating power for balanced two-sample *t*-test in Example 4.1.3

Parameters (Help)

Sample size   70

Effect size (Calculator)   0.3

Significance level   0.05

Power

Type of test   Two sample ⬍

H1   Greater ⬍

Power curve   No power curve ⬍

Note   t-test

Calculate

Output

```
Two-sample t-test

    n    d alpha  power
   70  0.3  0.05 0.5483

NOTE: n is number in *each* group
URL: http://psychstat.org/ttest
```

ticipants in group A and 40 participants in group B. Note that in the *Effect size* field, the input is 0.2:0.8:0.1, which generates a sequence of effect sizes from 0.2 to 0.8 with the interval 0.1. In the output, the power for each effect size is listed. The power curve (Figure 4.1.8) is displayed at the bottom of the output. The power curve can be used for interpolation. For example, to get a power at least 0.8, the effect size should be somewhere between 0.6 and 0.7.

Example 4.1.6: Calculate sample size given power and effect size in paired t-test

In a pre-test and post-test design, one may be interested in whether the post-test scores increase from pre-test scores (i.e., $H_1 : \mu_D > 0$). When the effect size in a paired *t*-test is known, we can determine how many participants are needed to attain the desired power level 0.8.

To conduct the power analysis, specify *Type of test* to be "Paired" and leave the *Sample size* field blank. Suppose the effect size is 0.4, the input and output for calculating the sample size are given in Figure 4.1.9. In the field of *Effect size*, input 0.4, and in the field of *power*, input 0.8. The default significance level 0.05 is used. By clicking the "**Calculate**"

Unbalanced two-sample t test

Parameters (Help)

Sample size of group 1	30
Sample size of group 2	40
Effect size (Calculator)	0.356
Significance level	0.05
Power	
H1	Two sided ⬍
Power curve	No power curve ⬍
Note	Unbalanced two-sample

Calculate

Output

```
Unbalanced two-sample t-test

   n1 n2     d alpha  power
   30 40 0.356  0.05 0.3065


NOTE: n1 and n2 are number in *each* group
URL: http://psychstat.org/ttest2n
```

button, we conclude that the required sample size is 40.

4.2   *Using R Package WebPower*

The power analysis conducted in the online interfaces can be done using the wp.t function in the R package WebPower. This function is adapted from the functions pwr.t.test and pwr.t2n.test from the R package pwr developed by Champely (2012). The detail of the function is:

```
wp.t(n1 = NULL, n2 = NULL, d=NULL, alpha = 0.05, power = NULL,
    type = c("two.sample", "one.sample", "paired", "two.sample.2n
    "), alternative = c("two.sided", "less", "greater"), tol = .
    Machine$double.eps^0.25)
```

The R input and output for the examples used above are given below:

```
> ## one sample t-test given sample size and effect size
> wp.t(n1=150, d=.2, type='one.sample')

One-sample t-test
```

n1: sample size for group 1
n2: sample size for group 2
d: effect size
alpha: significance level
power: statistical power
type: type of analysis. 1p, one sample; 2p, two sample with equal sample size; 2p2n, two sample with unequal sample size
alternative: alternative hypothesis
tol: tolerance in root solver.

Unbalanced two-sample t test

Figure 4.1.7: Input and output for power curve for unbalanced two-sample *t*-test in Example 4.1.5

Parameters (Help)

Sample size of group 1   30

Sample size of group 2   40

Effect size (Calculator)   0.2:0.8:0.1

Significance level   0.05

Power

H1   Two sided ⬍

Power curve   Show power curve ⬍

Note   Unbalanced two-sample

Calculate

Output

```
Unbalanced two-sample t-test

   n1 n2   d alpha  power
   30 40 0.2  0.05 0.1292
   30 40 0.3  0.05 0.2318
   30 40 0.4  0.05 0.3719
   30 40 0.5  0.05 0.5323
   30 40 0.6  0.05 0.6876
   30 40 0.7  0.05 0.8152
   30 40 0.8  0.05 0.9041


NOTE: n1 and n2 are number in *each* group
URL: http://psychstat.org/ttest2n
```

```
     n   d alpha    power
   150 0.2  0.05 0.682153

>
> ## paired t-test given sample size and effect size
> wp.t(n1=40, d=-.4, type='paired', alternative='less')

Paired t-test

    n    d alpha    power
   40 -0.4  0.05 0.7997378

NOTE: n is number of *pairs*

>
```

Figure 4.1.8: Power curve for unbalanced two-sample *t*-test in Example 4.1.5

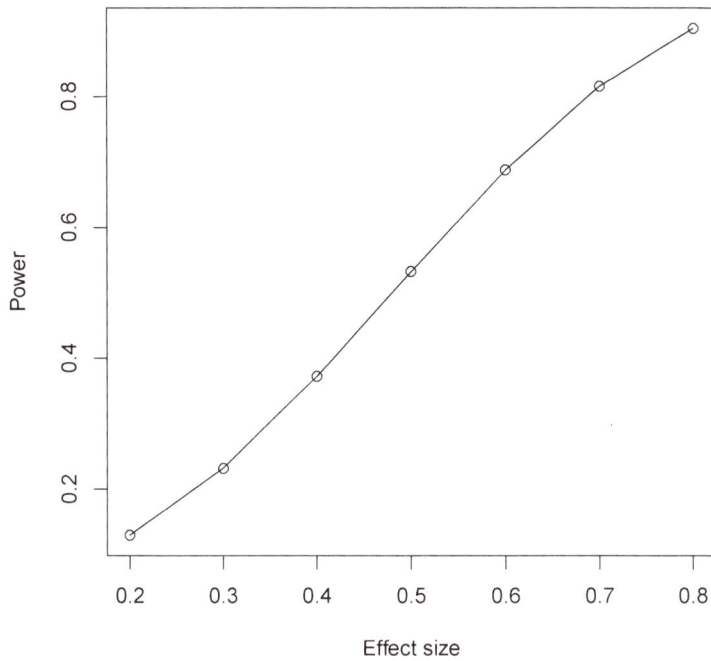

```
> ## paired t-test given power and effect size
> wp.t(power=.8, d=.4, type='paired', alternative='greater')

Paired t-test

          n   d alpha power
    40.02908 0.4  0.05   0.8

NOTE: n is number of *pairs*

>
> ## balanced two-sample t-test given sample size and effect size
> wp.t(n1=70, d=.3, alternative='greater')

Two-sample t-test

     n   d alpha     power
    70 0.3  0.05 0.5482577

NOTE: n is number in *each* group

>
> ## unbalanced two-sample t-test given sample size and effect
    size
```

One-sample, paired, two-sample balanced t test

Figure 4.1.9: Input and output for sample size planning for paired *t*-test in Example 4.1.6

Parameters (Help)

Sample size

Effect size (Calculator)   0.4

Significance level   0.05

Power   0.8

Type of test   Paired   ⬍

H1   Greater   ⬍

Power curve   No power curve   ⬍

Note   t-test

Calculate

Output

```
   Paired t-test

        n    d alpha power
    40.03 0.4  0.05   0.8


NOTE: n is number of *pairs*
URL: http://psychstat.org/ttest
```

```
> wp.t(n1=30, n2=40, d=.356, type='two.sample.2n')

Unbalanced two-sample t-test

   n1 n2     d alpha      power
   30 40 0.356  0.05 0.3064767

NOTE: n1 and n2 are number in *each* group

>
> ## unbalanced two-sample t-test given sample size and a
    sequence of effect sizes
> res <- wp.t(n1=30, n2=40, d=seq(.2,.8,.05), type='two.sample.2n
    ')
> res

Unbalanced two-sample t-test

   n1 n2    d alpha      power
   30 40 0.20  0.05 0.1291567
```

```
    30 40 0.25  0.05 0.1751916
    30 40 0.30  0.05 0.2317880
    30 40 0.35  0.05 0.2979681
    30 40 0.40  0.05 0.3719259
    30 40 0.45  0.05 0.4510800
    30 40 0.50  0.05 0.5322896
    30 40 0.55  0.05 0.6121937
    30 40 0.60  0.05 0.6876059
    30 40 0.65  0.05 0.7558815
    30 40 0.70  0.05 0.8151817
    30 40 0.75  0.05 0.8645929
    30 40 0.80  0.05 0.9040910


NOTE: n1 and n2 are number in *each* group


>
> ## generate a power curve
> plot(res, xvar='d', yvar='power')
>
> ## unbalanced two-sample t-test given sample size for one group
     , power and effect size
> wp.t(n1=50, power=.8, d=.5, type='two.sample.2n')


Unbalanced two-sample t-test

    n1       n2   d alpha power
    50 87.70891 0.5  0.05   0.8


NOTE: n1 and n2 are number in *each* group
```

4.3   Effect Size for t-tests

In WebPower, we use the statistic d (known as Cohen's d) as the measure
of effect size for power analysis in t-tests (Cohen, 1988, p. 20). For the
one-sample case, the population effect size is defined as

$$\delta = \frac{\mu - \mu_0}{\sigma},$$

where μ is the population mean of interest, μ_0 is the population value
under the null hypothesis, and σ is the population standard deviation.
The sample effect size is

$$d = \frac{\bar{y} - \mu_0}{s}$$

with $\bar{y}$ denoting the sample mean and s denoting the sample standard
deviation.

For matched pairs, the effect size is based on the difference scores
between all pairs, $y_{Di} = y_{1i} - y_{2i}$. The population effect size is

$$\delta = \frac{\mu_D}{\sigma_D} = \frac{\mu_1 - \mu_2}{\sigma_D},$$

where σ_D is the population standard deviation of difference scores. And the sample effect size is

$$d = \frac{\bar{y}_D}{s_D} = \frac{\bar{y}_1 - \bar{y}_2}{s_D},$$

where $\bar{y}_D$ denotes the sample mean of difference scores, and s_D denotes the sample standard deviation of difference scores.

For the independent two-sample t-test, the effect size is also called the standardized mean difference. The population effect size is defined as

$$\delta = \frac{\mu_1 - \mu_2}{\sigma}, \tag{4.3.1}$$

where μ_1 and μ_2 are the population means of two groups, and σ is the population standard deviation assumed to be equal across the two groups. δ is unknown but can be estimated by its sample effect size

$$d = \frac{\bar{y}_1 - \bar{y}_2}{s_p}, \tag{4.3.2}$$

where $\bar{y}_1$ and $\bar{y}_2$ denote the sample means of two groups. s_p is the unbiased estimator of the common variance (i.e., the square root of the pooled variance),

$$
\begin{aligned}
s_p &= \sqrt{\frac{(n_1 - 1)s_1^2 + (n_2 - 1)s_2^2}{n_1 + n_2 - 2}} \\
s_1^2 &= \frac{1}{n_1 - 1} \sum_{i=1}^{i=n_1} (y_{1i} - \bar{y}_1)^2, \\
s_2^2 &= \frac{1}{n_2 - 1} \sum_{i=1}^{i=n_2} (y_{2i} - \bar{y}_2)^2
\end{aligned}
$$

where n_1 and n_2 denote the sample sizes, and $s_1{}^2$ and $s_2{}^2$ denote the sample variances of the two groups, respectively. A special case is the balanced design where $n_1 = n_2 = n$, $s_p = \sqrt{(s_1^2 + s_2^2)/2}$.

4.3.1   Examples

Given the importance of obtaining an effect size, we show how to calculate an effect size under different conditions using several examples.

Suppose we have data observed in Table 4.3.1. The null hypothesis is $H_0 : \mu = 0$. From the data, we have the sample mean $\bar{y} = 19.4$. The sample standard deviation is

Example 4.3.1: Effect size of one-sample t-test

$$s = \sqrt{\frac{1}{5-1}[(22 - 19.4)^2 + (12 - 19.4)^2 + (25 - 19.4)^2 + (18 - 19.4)^2 + (20 - 19.4)^2]} = 4.879.$$

Therefore, the sample effect size is $d = \bar{y}/s = 19.4/4.879 = 3.976$. We might make a hypothesis that the population effect size is 3.976. The method of evaluating power by substituting the population effect size with its sample estimate is called post hoc power, and the results can be misleading regardless of how large the sample size is (Yuan & Maxwell, 2005).

	Scores
	22
	12
	25
	18
	20
Mean	19.4

Table 4.3.1: Example data for effect size calculation of one-sample t-test

Suppose we have the data shown in Table 4.3.2. Based on the pre-test scores and post-test scores, we can calculate the difference scores. The mean of the difference scores is $\bar{y}_D = 2$. The sample standard deviation of the difference scores is

Example 4.3.2: Effect size of paired t-test

$$s_D = \sqrt{\frac{1}{5-1}[(3-2)^2 + (7-2)^2 + (0-2)^2 + (1-2)^2 + (-1-2)^2]} = 3.162.$$

Therefore, the sample effect size is $d = \bar{y}_D / s_D = 2/3.162 = 0.633$.

	Pre-test score	Post-test score	Difference score
	22	25	3
	12	19	7
	25	25	0
	18	19	1
	20	19	-1
Mean	19.4	21.4	2

Table 4.3.2: Example data for effect size calculation of paired t-test

Suppose we have the data shown in Table 4.3.3. From the data, we have the pooled variance $s_p^2 = (s_1^2 + s_2^2)/2 = 17.3$. The sample effect size is $d = (\bar{y}_1 - \bar{y}_2)/s_p = (21.4 - 19.4)/\sqrt{17.3} = 0.481$.

Example 4.3.3: Effect size of balanced two-sample t-test

	Group A scores	Group B scores
	22	25
	12	19
	25	25
	18	19
	20	19
Mean	19.4	21.4
Variance	23.8	10.8

Table 4.3.3: Example data for effect size calculation of balanced two-sample t-test

Example 4.3.4: Effect size of unbalanced two-sample t-test

Suppose we have data shown in Table 4.3.4. From the data, we have the pooled variance $s_p^2 = [(6-1)9.467 + (9-1)13.278]/(6+9-2) = 11.812$. The sample effect size is $d = (\bar{y}_1 - \bar{y}_2)/s_p = 1.223/\sqrt{11.812} = 0.356$.

	Group A scores	Group B scores
	13	20
	20	15
	14	16
	19	15
	18	16
	20	23
		20
		17
		25
Mean	17.333	18.556
Variance	9.467	13.278

Table 4.3.4: Example data for effect size calculation of unbalanced two-sample t-test

4.3.2   *Effect size calculator*

WebPower provides an online calculator to calculate the effect size for the one-sample t-test, paired t-test, and two-sample t-test based on the user-provided information. The interface for the calculator is shown in Figure 4.3.1, which can be brought up by clicking the link "**Calculator**" in the power calculation interface.

http://psychstat.org/ttesteffect

For the one-sample t-test, one can provide the population mean for null and alternative hypotheses as well as the standard deviation to get the effect size. For the paired two-sample t-test, one would need to provide the mean and standard deviation of the difference. For the two-sample t-test, the means of the two groups and the common standard deviation are required.

Effect size can also be estimated from individual/empirical data. In this case, a registered user can upload a data set. For the one-sample t-test, the data set should include a single column of data. For the paired two-sample t-test, the data set should have two columns with each row including data for a pair. For the two-sample independent t-test, the data set also includes two columns but with the first column representing data for both groups and the second column a grouping variable. Some sample data are given in Table 4.3.5.

As an example, we saved the data for the two-sample example in Table 4.3.5 into a file (http://psychstat.org/tdata). Using the data, we obtained the results shown in Figure 4.3.2. Note that in addition to the output of the effect size, a t-test was also conducted for the data and the output of the t-test was provided.

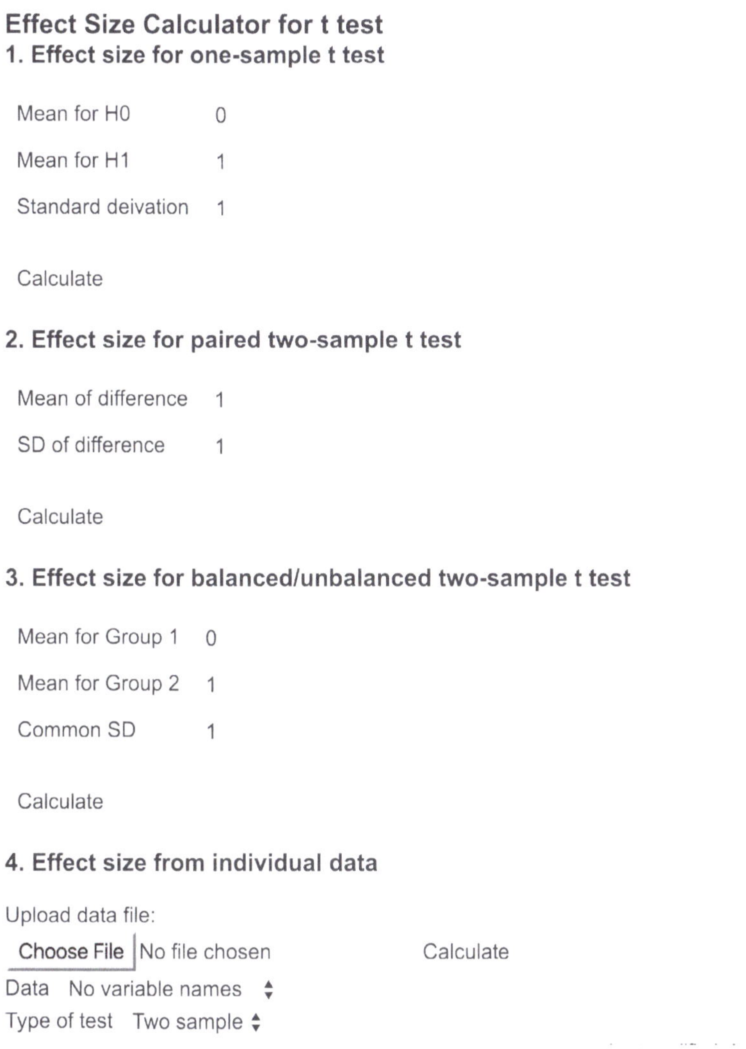

Effect Size Calculator for t test
1. Effect size for one-sample t test

Figure 4.3.1: Effect size calculator for t-test power analysis

Mean for H0   0

Mean for H1   1

Standard deivation   1

Calculate

2. Effect size for paired two-sample t test

Mean of difference   1

SD of difference   1

Calculate

3. Effect size for balanced/unbalanced two-sample t test

Mean for Group 1   0

Mean for Group 2   1

Common SD   1

Calculate

4. Effect size from individual data

Upload data file:

Choose File │ No file chosen   Calculate

Data   No variable names ⬍
Type of test   Two sample ⬍

4.4 *Technical Details*

4.4.1 *One-sample t-test*

In a one-sample t-test, we are interested in whether the population mean μ is different from a specific value μ_0 (usually $\mu_0 = 0$). Thus the null hypothesis is

$$H_0 : \mu = \mu_0.$$

The alternative hypothesis can be either two-sided or one-sided to indicate the difference:

$$H_{11} : \mu \neq \mu_0,$$

or

$$H_{12} : \mu > \mu_0,$$

one-sample	paired two-sample		two-sample	
y	y1	y2	y	group
22	22	25	13	0
12	12	19	20	0
25	25	25	14	0
18	18	19	19	0
20	20	19	18	0
			20	0
			20	1
			15	1
			16	1
			15	1
			16	1
			23	1
			20	1
			17	1
			25	1

Table 4.3.5: Format of data files for effect size calculator for t-test

or

$$H_{13}: \mu < \mu_0.$$

Let $\bar{y}$ denote the sample mean and s^2 denote the sample variance. Assume that $y_i \sim N(\mu, \sigma^2)$. The corresponding t test statistic given the sample size n is

$$
\begin{aligned}
t = \frac{\bar{y} - \mu_0}{s/\sqrt{n}} &= \frac{(\bar{y} - \mu) + (\mu - \mu_0)}{s/\sqrt{n}} \\
&= \frac{[(\bar{y} - \mu) + (\mu - \mu_0)]/\sqrt{\sigma^2/n}}{\sqrt{s^2/\sigma^2}} \\
&= \frac{(\bar{y} - \mu)/\sqrt{\sigma^2/n} + (\mu - \mu_0)/\sqrt{\sigma^2/n}}{\sqrt{\frac{(n-1)s^2}{\sigma^2(n-1)}}}.
\end{aligned}
\tag{4.4.1}
$$

Because $y_i \sim N(\mu, \sigma^2)$, $(\bar{y} - \mu)/\sqrt{\sigma^2/n} \sim N(0,1)$, $\frac{(n-1)s^2}{\sigma^2} \sim \chi^2(n-1)$, and $\bar{y} - \mu_0$ and s are independent, the t statistic in Equation 4.4.1 follows a t distribution $t(n-1, (\mu - \mu_0)/\sqrt{\sigma^2/n})$. The non-centrality parameter is

$$\lambda = \frac{\mu - \mu_0}{\sqrt{\sigma^2/n}} = \sqrt{n}\delta,$$

with $\delta = (\mu - \mu_0)/\sigma$ denoting the population effect size.

Under the null hypothesis, $\lambda = \frac{\mu - \mu_0}{\sqrt{\sigma^2/n}} = 0$. The t statistic follows a Student's t distribution with degrees of freedom $n - 1$. If the t statistic is larger than the critical value $t_{1-\alpha/2}$ or smaller than $t_{\alpha/2}$ in a two-sided test, or larger than $t_{1-\alpha}$ or smaller than t_{α} in a one-sided tests, the null hypothesis H_0 is rejected.

Effect Size Calculator for t test

4. Effect size from individual data

Upload data file:

| Choose File | tdata.txt   Calculate

Data   With variable names ↕
Type of test   Two sample ↕

Effect size output

The effect size for independent two-sample t test = **0.3556**
The sample mean of the two groups = **17.3333 18.5556**
The sample SD of the the two groups = **3.0768 3.6439**
The pooled SD of the the two groups = **3.4369**

Output from independent two-sample t-test

```
        Welch Two Sample t-test

data:  y by group
t = -0.69949, df = 12.107, p-value = 0.4975
alternative hypothesis: true difference in means is not equal to 0
95 percent confidence interval:
 -5.025568  2.581124
sample estimates:
mean in group 0 mean in group 1
      17.33333        18.55556
```

Under the alternative hypothesis, the t statistic in Equation 4.4.1 follows a non-central t distribution $t(n-1, \frac{\mu-\mu_0}{\sqrt{\sigma^2/n}})$. Power can be calculated from this distribution as

$$\pi = \begin{cases} 1 - t_{n-1,\lambda}(t^{-1}_{n-1,1-\alpha/2}) + t_{n-1,\lambda}(t^{-1}_{n-1,\alpha/2}) & H_{11} \\ 1 - t_{n-1,\lambda}(t^{-1}_{n-1,1-\alpha}) & H_{12} \\ t_{n-1,\lambda}(t^{-1}_{n-1,\alpha}) & H_{13} \end{cases} \quad (4.4.2)$$

where $t^{-1}_{n-1,1-\alpha}$ is the critical value of a Student's t distribution given the probability $1 - \alpha$, and $t_{n-1,\lambda}$ is the cumulative non-central t distribution function.

Since usually we do not know the population mean μ and population variance σ^2, the estimated $\hat{\lambda}$ and the sample effect size d are used to calculate power

$$\hat{\lambda} = d\sqrt{n} \quad = \quad \frac{\bar{y} - \mu_0}{s}\sqrt{n}.$$

Again, if the population effect size is actually different from the sample effect size, the power analysis results can be very misleading.

4.4.2   *Two-sample t-test*

A two-sample t test is used to determine whether two independent population means are equal. Thus the null hypothesis is

$$H_0 : \mu_1 - \mu_2 = 0.$$

The alternative hypothesis can be either two-sided or one-sided to indicate the difference:

$$H_{11} : \mu_1 - \mu_2 \neq 0,$$

or

$$H_{12} : \mu_1 - \mu_2 > 0,$$

or

$$H_{13} : \mu_1 - \mu_2 < 0.$$

Let $\bar{y}_1$ and $\bar{y}_2$ denote the sample means and $s_1{}^2$ and $s_2{}^2$ denote the sample variances of the two groups. Assume that $y_{1i} \sim N(\mu_1, \sigma^2)$ and $y_{2i} \sim N(\mu_2, \sigma^2)$. The corresponding t test statistic given the sample sizes n_1 and n_2 is

$$
\begin{aligned}
t &= \frac{\bar{y}_1 - \bar{y}_2}{s_p \sqrt{\frac{1}{n_1} + \frac{1}{n_2}}} = \frac{\bar{y}_1 - \bar{y}_2 - (\mu_1 - \mu_2) + (\mu_1 - \mu_2)}{s_p \sqrt{\frac{1}{n_1} + \frac{1}{n_2}}} \\[2ex]
&= \frac{\frac{(\bar{y}_1 - \bar{y}_2) - (\mu_1 - \mu_2)}{\sigma \sqrt{\frac{1}{n_1} + \frac{1}{n_2}}} + \frac{(\mu_1 - \mu_2)}{\sigma \sqrt{\frac{1}{n_1} + \frac{1}{n_2}}}}{\sqrt{\frac{s_p^2}{\sigma^2}}} \\[2ex]
&= \frac{\frac{(\bar{y}_1 - \bar{y}_2) - (\mu_1 - \mu_2)}{\sigma \sqrt{\frac{1}{n_1} + \frac{1}{n_2}}} + \frac{(\mu_1 - \mu_2)}{\sigma \sqrt{\frac{1}{n_1} + \frac{1}{n_2}}}}{\sqrt{\frac{(n_1 + n_2 - 2)s_p^2}{\sigma^2 (n_1 + n_2 - 2)}}},
\end{aligned}
\tag{4.4.3}
$$

where s_p is an unbiased estimator of the common variance,

$$s_p = \sqrt{\frac{(n_1 - 1)s_1^2 + (n_2 - 1)s_2^2}{n_1 + n_2 - 2}}.$$

With $\frac{(\bar{y}_1 - \bar{y}_2) - (\mu_1 - \mu_2)}{\sigma \sqrt{\frac{1}{n_1} + \frac{1}{n_2}}} \sim N(0,1)$ and $\frac{(n_1 + n_2 - 2)s_p^2}{\sigma^2} \sim \chi^2(n_1 + n_2 - 2)$, the t statistic in Equation 4.4.3 follows a non-central t distribution $t(n_1 + n_2 - 2, \frac{(\mu_1 - \mu_2)}{\sigma \sqrt{\frac{1}{n_1} + \frac{1}{n_2}}})$. The non-centrality parameter is

$$\lambda = \frac{\mu_1 - \mu_2}{\sigma \sqrt{\frac{1}{n_1} + \frac{1}{n_2}}} = \sqrt{\frac{n_1 n_2}{n_1 + n_2}} \delta,$$

with $\delta = \frac{\mu_1 - \mu_2}{\sigma}$ denoting the population effect size.

A special case is the balanced design where $n_1 = n_2 = n$,

$$s_p = \sqrt{\frac{s_1^2 + s_2^2}{2}},$$

$$t = \frac{\bar{y}_1 - \bar{y}_2}{\sqrt{\frac{s_1^2 + s_2^2}{n}}},$$

$$\lambda = \sqrt{\frac{n}{2}}\delta.$$

Under the null hypothesis, $\lambda = \frac{(\mu_1 - \mu_2)}{\sigma\sqrt{\frac{1}{n_1} + \frac{1}{n_2}}} = 0$. The t statistic follows a Student's t-distribution with degrees of freedom $n_1 + n_2 - 2$. Otherwise, it follows a noncentral t-distribution. Under the alternative hypothesis, power can be calculated as

$$\pi = \begin{cases} 1 - t_{n_1+n_2-2,\lambda}\left(t_{n-1,1-\alpha/2}^{-1}\right) + t_{n_1+n_2-2,\lambda}\left(t_{n_1+n_2-2,\alpha/2}^{-1}\right) & H_{11} \\ 1 - t_{n_1+n_2-2,\lambda}\left(t_{n_1+n_2-2,1-\alpha}^{-1}\right) & H_{12} \\ t_{n_1+n_2-2,\lambda}\left(t_{n-1,\alpha}^{-1}\right) & H_{13} \end{cases}.$$

In practice, the estimated $\hat{\lambda}$ and sample effect size d are used to calculate power by assuming the sample effect size is the same as the population effect size:

$$\hat{\lambda} = d\sqrt{\frac{n_1 n_2}{n_1 + n_2}} = \frac{(\bar{y}_1 - \bar{y}_2)}{s_p}\sqrt{\frac{n_1 n_2}{n_1 + n_2}}.$$

4.4.3   Paired t-test

The paired t-test is used to test whether the matched pairs have equal means. The difference scores between all pairs are calculated by $y_{Di} = y_{1i} - y_{2i}$ (Cohen, 1988). The null hypothesis is

$$H_0 : \mu_D = \mu_1 - \mu_2 = 0.$$

The alternative hypothesis can be either two-sided or one-sided to indicate the difference:

$$H_{11} : \mu_D \neq 0,$$

or

$$H_{12} : \mu_D > 0,$$

or

$$H_{13} : \mu_D < 0.$$

Let $\bar{y}_D$ denote the sample means of the difference scores, and s_D^2 denote the sample variance of the difference scores. Assume that

$y_{1i} - y_{2i} \sim N(\mu, \sigma_D^2)$. The corresponding t test statistic given the number of pairs n is

$$
\begin{aligned}
t = \frac{\bar{y}_D - 0}{s_D/\sqrt{n}} &= \frac{\bar{y}_D - \mu_D + \mu_D}{s_D/\sqrt{n}} \\
&= \frac{[(\bar{y}_D - \mu_D) + \mu_D]/\sqrt{\frac{\sigma_D^2}{n}}}{\sqrt{\frac{s_D^2}{\sigma_D^2}}} \\
&= \frac{(\bar{y}_D - \mu_D)/\sqrt{\frac{\sigma_D^2}{n}} + \mu_D/\sqrt{\frac{\sigma_D^2}{n}}}{\sqrt{\frac{(n-1)s_D^2}{\sigma_D^2(n-1)}}},
\end{aligned}
\tag{4.4.4}
$$

where $\sigma_D^2 = \sigma_{y1-y2}^2 = \sigma_1^2 + \sigma_2^2 - 2\rho\sigma_1\sigma_2$, $s_D^2 = s_{1-2}^2 = s_1^2 + s_2^2 - 2rs_1s_2$. Using the assumption of equal variance $\sigma_{y1}^2 = \sigma_{y2}^2 = \sigma^2$, $\sigma_D^2 = 2\sigma^2(1 - \rho)$.

The non-centrality parameter of the non-central t distribution is

$$
\lambda = \frac{\mu_D}{\sqrt{\frac{\sigma_D^2}{n}}} = \frac{\mu_1 - \mu_2}{\sqrt{\frac{\sigma_D^2}{n}}} = \sqrt{n}\delta,
$$

with $\delta = (\mu_1 - \mu_2)/\sigma_D^2$ denoting the population effect size. Also $\hat{\lambda} = \frac{\bar{y}_D}{\sqrt{\frac{s_D^2}{n}}} = \sqrt{n}d$. Under the alternative hypothesis, the power function is same as the one used in one sample t-test 4.4.2.

4.5   Exercises

1. A researcher believes that the mean of a test is higher than 80. If he collects data from 25 participants with mean and variance 100 and 30 respectively, what are the null hypothesis (H_0) and the alternative hypothesis (H_1)? What is the statistical power when the alpha level is set at 0.05?

2. The test scores from two classes with different textbooks are recorded. If each class has 25 students, what is the sample effect size?

	Class 1	Class 2
Mean	100	125
Variance	30	35

3. Using the same information in Exercise 2, what is the power for 25 participants per class at the alpha level 0.1 if assuming the population effect size is the same as the sample effect size?

4. If the mean of a test of using A textbook is 120, the mean of using B textbook is 140, and the common variance is 1225. Generate a power

curve with the sample size per group ranging from 10 to 100 with an interval of 10. What is the total sample size needed for a balanced design to get a power 0.9 at the alpha level 0.05?

5   *Statistical Power Analysis for One-Way ANOVA*

Zhiyong Zhang
Department of Psychology
University of Notre Dame

One-way analysis of variance (one-way ANOVA) is a technique used to compare means of two or more groups (e.g., Maxwell & Delaney, 2004). The ANOVA tests the null hypothesis that samples in two or more groups are drawn from populations with the same mean values. The ANOVA analysis typically produces an F-statistic, the ratio of the sample variance of the between-group to that of the within-group. If the group means are drawn from populations with the same mean values, the variance of the group means should not be too high when compared with the variance within the groups. A higher ratio, therefore, implies that the samples were drawn from populations with different mean values.

5.1   *How to Conduct Power Analysis for One-way ANOVA*

The primary software interface for power analysis for one-way ANOVA is shown in Figure 5.1.1.. Within the interface, a user can supply different parameter values and select different options for power analysis. Among the five parameters, *Number of groups*[1], *Sample size*, *Effect size*, *Significance level*, and *Power*, one and only one can be left blank.

http://psychstat.org/anova

[1] Also called the number of levels

- The *Number of groups* tells how many groups are involved in the study design.

- The *Sample size* is the **total** number of participants from all groups. For example, if the sample size for each group is 25 for 4 groups, the total sample size is 100. Multiple sample sizes can be provided in two ways to calculate power for each sample size. First, multiple sample sizes can be supplied and separated by white spaces, e.g., 100 150 200 will calculate power for the three sample sizes 100, 150

One-way ANOVA

Parameters (Help)

Number of groups	4
Sample size	100
Effect size (Calculator)	0.5
Significance level	0.05
Power	
Type of analysis	Overall ↕
Power curve	No power curve ↕
Note	One-way ANOVA

Calculate

Figure 5.1.1: Software interface of power analysis for one-way ANOVA

and 200. Second, a sequence of sample sizes can be generated using the method `s:e:i` with `s` denoting the starting sample size, `e` as the ending sample size, and `i` as the interval. Note the values are separated by colon ":". For example, 100:150:10 will generate a sequence of sample sizes - 100 110 120 130 140 150. The default sample size, as shown in Figure 5.1.1, is 100.

- The *Effect size*[2] specifies the population group difference. Multiple effect sizes or a sequence of effect sizes can also be supplied using the same method for sample size. By default, the value is 0.5. Determining the effect size is critical but not trivial work. To help a user obtain effect sizes, a calculator has been developed and can be used by clicking the link "**Calculator**".

[2] More on effect size will be provided in Section 5.3

- The *Significance level*[3] for power calculation is needed but usually set at the default value 0.05.

[3] Type I error rate or alpha level

- The *Power* specifies the desired statistical power.

- The software can calculate power for overall effect and for a specific contrast. One can choose the type of analysis through the option "*Type of analysis*".

- In addition to the required input, one can also request the plot of a power curve if multiple sample sizes or effect sizes are provided.

- A note (less than 200 characters) can be provided to save basic information on the analysis for future reference for registered users.

5.1.1 Examples

A student hypothesizes that freshman, sophomore, junior and senior college students have different attitudes towards obtaining arts degrees. Based on his prior knowledge, he expects that the population effect size is about 0.25. If he plans to interview 25 students on their attitude in each student group, what is the power for him to find the significant difference among the four groups?

The input and output for calculating power for this study are given in Figure 5.1.2. In the field of *Number of groups*, input 4, the total number of groups; in the field of *Sample size*, input 100 (= 25 × 4), the total sample size of the four groups; and in the field of *Effect size*, input 0.25, the expected effect size. The default significance level 0.05 is used although one can change it to a different value. The field for *Power* is left blank because it will be calculated. A simple note "One-way ANOVA" is also added in the *Note* field. By clicking the "**Calculate**" button, the statistical power is given in the output immediately. For the current design, the power is 0.5182.

Example 5.1.1: Calculate power given sample size and effect size

One-way ANOVA

Parameters (Help)

Number of groups	4
Sample size	100
Effect size (Calculator)	0.25
Significance level	0.05
Power	
Type of analysis	Overall ↕
Power curve	No power curve ↕
Note	One-way ANOVA

Calculate

Output

```
Power for One-way ANOVA

   k   n    f alpha  power
   4 100 0.25  0.05 0.5182


NOTE: n is the total sample size (overall)
URL: http://psychstat.org/anova
```

Figure 5.1.2: Input and output for calculating power for one-way ANOVA in Example 5.1.1

A power curve is a line plot of statistical power along with given sample sizes. In Example 5.1.1, the power is 0.518 with the sample size 25 in each group. What is the power for a different sample size, say, 50 in each group? One can investigate the power of different sample sizes and plot a power curve.

The input and output for calculating power for the study in Example 5.1.1 with a sample size from 100 to 200 (25 to 50 each group) with an interval of 20 are given in Figure 5.1.3. Note that in the *Sample size* field, the input is 100:200:20. In the output, the power for each sample size from 100 to 200 with the interval 20 is listed. Especially, with the sample size 180 (the sample size is 45 for each group), the power is about 0.804. In the input, we also choose *"Show power curve"* for the *Power curve* field. In the output, the power curve (Figure 5.1.4) is displayed at the bottom of the output. The power curve can be used for interpolation. For example, to get a power 0.8, about 45 students are needed for each group.

Example 5.1.2: Power curve

In practice, a power 0.8 is often desired. Given the power, the sample size can also be calculated as shown in Figure 5.1.5. In this situation, the *Sample size* field is left blank while in the *Power* field, the value 0.8 is the input. In the output, we can see a sample size 179 is needed to obtain a power 0.8, that is about 45 for each group.

Example 5.1.3: Calculate sample size given power and effect size

One can also calculate the minimum effect to achieve certain power given a sample size. As shown in Figure 5.1.6, we leave the *Effect size* field blank but provide information on *Sample size* (100) and *Power* (0.8). In the output, we can see that the obtained effect size is 0.337. It means that to get a power 0.8 with the sample size 100, the population effect size has to be at least 0.337.

Example 5.1.4: Minimum detectable effect

Suppose that in addition to the overall effect, the student is especially interested in the difference between freshman and senior college students. He hypothesizes that the effect size would be 0.25. To calculate power for a total sample size 100, input 0.25 as the effect size (as shown in Figure 5.1.7). Then, set the *Type of analysis* to be "Contrast, two-sided". In the output, the power is 0.6967 for detecting the difference of an effect of 0.25 between freshman and senior college students.

Example 5.1.5: Calculate power for a given contrast

5.2 *Using R WebPower for Power Analysis for One-way ANOVA*

The online power analysis is carried out using the R package WebPower on a Web server. The package can be directly used within R for power

One-way ANOVA

Figure 5.1.3: Input and output for power curve for one-way ANOVA in Example 5.1.2

Parameters (Help)

Number of groups	4
Sample size	100:200:20
Effect size (Calculator)	0.25
Significance level	0.05
Power	
Type of analysis	Overall ⬍
Power curve	Show power curve ⬍
Note	One-way ANOVA

Calculate

Output

```
Power for One-way ANOVA

    k   n     f alpha  power
    4 100  0.25  0.05 0.5182
    4 120  0.25  0.05 0.6065
    4 140  0.25  0.05 0.6837
    4 160  0.25  0.05 0.7494
    4 180  0.25  0.05 0.8040
    4 200  0.25  0.05 0.8485


NOTE: n is the total sample size (overall)
URL: http://psychstat.org/anova
```

analysis for one-way ANOVA. Specifically, the function wp.anova is used. This function is adapted from the function pwr.anova.test from the R package pwr developed by Champely (2012). In addition to power analysis for the main effect, we also improved the function to calculate power for a contrast. The basic usage of the function is provided below.

```
wp.anova(k = NULL, n = NULL, f = NULL, alpha = 0.05, power = NULL
    , type = c("overall", "two.sided", "greater", "less"))
```

f: effect size
k: number of groups
n: sample size
alpha: significance level
power: statistical power
type: overall, contrast (two.sided, greater, less)

For example, the R input and output for Example 5.1.1 are given below. When the sample size is provided, the power will be calculated.

```
> wp.anova(f=0.25,k=4,n=100,alpha=0.05)

Power for One-way ANOVA

    k   n     f alpha      power
```

Figure 5.1.4: Power curve for one-way ANOVA in Example 5.1.2

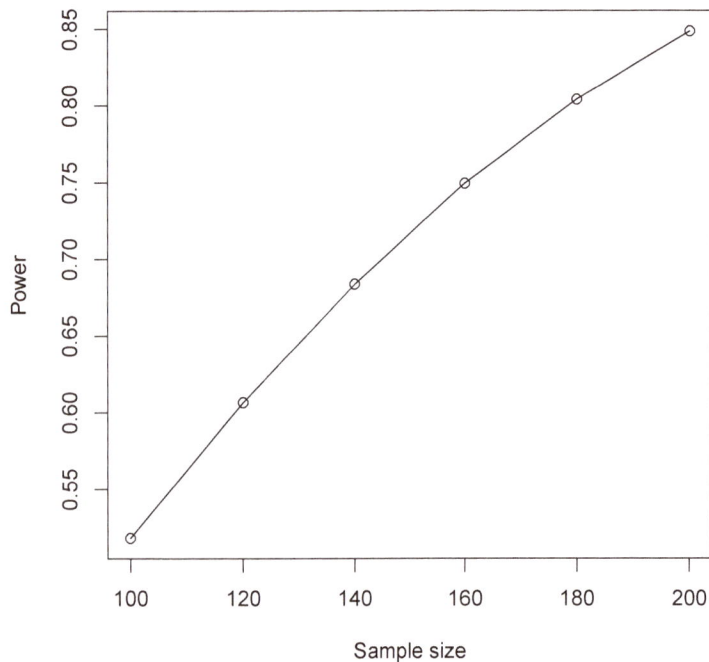

Figure 5.1.4: Power curve for one-way ANOVA in Example 5.1.2

```
     4 100 0.25  0.05 0.5181755

NOTE: n is the total sample size adding all groups (overall)
WebPower URL: http://psychstat.org/anova
```

The R input and output for generating the power curve in Example 5.1.2 are given below.

```
> example <- wp.anova(f=0.25,k=4,n=seq(100,200,10),alpha=0.05)
> example

Power for One-way ANOVA

    k   n    f alpha     power
    4 100 0.25  0.05 0.5181755
    4 110 0.25  0.05 0.5636701
    4 120 0.25  0.05 0.6065228
    4 130 0.25  0.05 0.6465721
    4 140 0.25  0.05 0.6837365
    4 150 0.25  0.05 0.7180010
    4 160 0.25  0.05 0.7494045
    4 170 0.25  0.05 0.7780286
    4 180 0.25  0.05 0.8039869
    4 190 0.25  0.05 0.8274169
    4 200 0.25  0.05 0.8484718
```

seq(100,200,10) generates a sequence of sample sizes.
The function plot generates a power curve similar to the one in Figure 5.1.4.

One-way ANOVA

Parameters (Help)

Number of groups	4
Sample size	
Effect size (Calculator)	0.25
Significance level	0.05
Power	0.8
Type of analysis	Overall ⬍
Power curve	Show power curve ⬍
Note	One-way ANOVA

Calculate

Output

```
Power for One-way ANOVA

    k     n     f alpha power
    4 178.4 0.25  0.05   0.8


NOTE: n is the total sample size (overall)
URL: http://psychstat.org/anova
```

```
NOTE: n is the total sample size adding all groups (overall)
WebPower URL: http://psychstat.org/anova

> plot(example, type='b')
```

Furthermore, we can also use R to estimate both the sample size and the minimum detectable effect size.

```
> wp.anova(f=0.25,k=4,n=NULL,alpha=0.05,power=0.8)

Power for One-way ANOVA

    k        n     f alpha power
    4 178.3971 0.25  0.05   0.8

NOTE: n is the total sample size adding all groups (overall)
WebPower URL: http://psychstat.org/anova

> wp.anova(f=NULL,k=4,n=100,alpha=0.05,power=0.8)

Power for One-way ANOVA
```

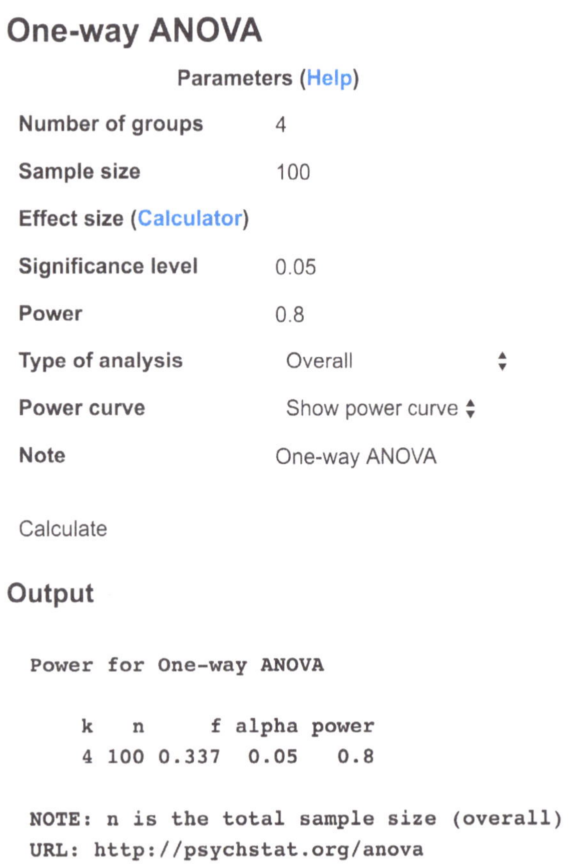

One-way ANOVA

Figure 5.1.6: Input and output for minimum effect size determination for one-way ANOVA in Example 5.1.4

Parameters (Help)

Number of groups	4
Sample size	100
Effect size (Calculator)	
Significance level	0.05
Power	0.8
Type of analysis	Overall ⬍
Power curve	Show power curve ⬍
Note	One-way ANOVA

Calculate

Output

```
Power for One-way ANOVA

    k    n       f alpha power
    4  100  0.337   0.05    0.8


NOTE: n is the total sample size (overall)
URL: http://psychstat.org/anova
```

```
    k    n          f alpha power
    4  100  0.3369881   0.05    0.8


NOTE: n is the total sample size adding all groups (overall)
WebPower URL: http://psychstat.org/anova
```

Finally, to conduct power analysis for a contrast, the input and output are shown below.

```
> wp.anova(f=.25, k=4, n=100, type='two.sided')

Power for One-way ANOVA

    k    n     f alpha      power
    4  100  0.25  0.05  0.6967142


NOTE: n is the total sample size adding all groups (contrast, 2
    side)
WebPower URL: http://psychstat.org/anova
```

One-way ANOVA

Figure 5.1.7: Input and output for calculating power of a contrast for one-way ANOVA in Example 5.1.5

Parameters (Help)

Number of groups	4
Sample size	100
Effect size (Calculator)	0.25
Significance level	0.05
Power	
Type of analysis	Contrast, two-sided ↕
Power curve	Show power curve ↕
Note	One-way ANOVA

Calculate

Output

```
Power for One-way ANOVA

   k   n    f alpha  power
   4 100 0.25  0.05 0.6967


NOTE: n is the total sample size (contrast, two.sided)
URL: http://psychstat.org/anova
```

5.3   *Effect Size for One-way ANOVA*

We use the statistic f as the measure of effect size for one-way ANOVA as in Cohen (1988, p. 275). The f is the ratio of the between-group standard deviation of the effect to be tested σ_b (or the standard deviation of the group means, or between-group standard deviation) and the common within-group standard deviation σ_w:

$$f = \frac{\sigma_b}{\sigma_w}.$$

Given the two quantities σ_b and σ_w, the effect size can be determined. Cohen defined the size of effect as: small 0.1, medium 0.25, and large 0.4.

The effect size can be determined based on the literature review or empirical data. In doing so, we assume that the population effect size is the same as the sample or empirical effect size, which may not be true. For example, based on the literature, one can obtain the information for each group in ANOVA as shown below:

Group	1	2	...	G
Group size	n_1	n_2	...	n_G
Mean	m_1	m_2	...	m_G
Variance	s_1^2	s_2^2	...	s_G^2

G: number of groups
n_g: sample size of group g
m_g: mean of group g
s_g^2: variance of group g

With such information, the between group standard deviation σ_b can be calculated by

$$\sigma_b = \sqrt{\sum_{g=1}^{G} w_g (m_g - \bar{m})^2}$$

with

$$\bar{m} = \sum_{g=1}^{G} w_g m_g$$

where w_g is the weight

$$w_g = \frac{n_g}{\sum_{i=1}^{G} n_g}.$$

A special case is for the balanced design where $n_g = n$ which leads to $w_g = 1/G$ and

$$\sigma_b = \sqrt{\sum_{g=1}^{G} (m_g - \bar{m})^2/G,}$$

$$\bar{m} = \sum_{g=1}^{G} m_g/G.$$

Therefore, the effect size can be calculated for both balanced and unbalanced designs. For the within-group standard deviation, it is calculated as

$$\sigma_w = \sqrt{\frac{\sum_{g=1}^{G} s_g^2}{G}}.$$

5.3.1   Effect size for a specific contrast (comparing two selected groups)

In addition to the power for the global F test, one can also conduct power analysis for a specific contrast. In this case, the effect size has to be calculated differently. Suppose we are interested in the difference between two groups i and j. Then the effect size is

$$f = \frac{\sigma_b}{\sigma_w} = \frac{\sqrt{(m_i - m_j)^2/(1/w_i + 1/w_j)}}{\sigma_w}.$$

5.3.2   Examples

To show how to calculate the effect sizes under different conditions, consider the following examples. Suppose the data are presented as below.

	Group	Freshman	Sophomore	Junior	Senior
	Group size	25	25	25	25
Balanced	Mean	2.0	3.0	3.6	4
	Variance	9	9	9	9
	Group size	10	20	30	40
Unbalanced	Mean	2.0	3.0	3.6	4
	Variance	9	9	9	9

Table 5.3.1: Example data for effect size calculation

Example 5.3.1: Overall effect size of the balanced design

We first calculate the overall effect size for the balanced design. From the data, we have the weight $w_g = w = 1/4$. The grand mean is $\bar{m} = 1/4 \times (2 + 3 + 3.6 + 4) = 3.15$. The between group variance is $\sigma_b^2 = \frac{1}{4}[(2-3.15)^2 + (3-3.15)^2 + (3.6-3.15)^2 + (4-3.15)^2] = .5675$. For the within group variance, $\sigma_w^2 = (9+9+9+9)/4 = 9$. Therefore, the effect size is $f = \sqrt{\sigma_b^2/\sigma_w^2} = \sqrt{.5675/9} = .251$.

Example 5.3.2: Overall effect size of the unbalanced design

We then calculate the overall effect size for the unbalanced design. From the data, we have the weight $w_1 = 0.1, w_2 = 0.2, w_3 = 0.4, w_4 = 0.4$. The grand mean is $\bar{m} = .1 \times 2 + .2 \times 3 + .3 \times 3.6 + .4 \times 4 = 3.48$. The between group variance is $\sigma_b^2 = .1 \times (2-3.48)^2 + .2 \times (3-3.48)^2 + .3 \times (3.6-3.48)^2 + .4 \times (4-3.48)^2] = .3776$. For the within group variance, $\sigma_w^2 = (9+9+9+9)/4 = 9$. Therefore, the effect size is $f = \sqrt{\sigma_b^2/\sigma_w^2} = \sqrt{.3776/9} = .205$.

Example 5.3.3: Effect size for contrast freshman and senior of the balanced design

We now calculate the effect size for comparing freshman and senior students in the balanced design. From the data, we have the weight $w_g = w = 1/4$. The between group variance is $\sigma_b^2 = (4-2)^2/(1/.25 + 1/.25) = 0.5$. For the within group variance, $\sigma_w^2 = (9+9+9+9)/4 = 9$. Therefore, the effect size is $f = \sqrt{\sigma_b^2/\sigma_w^2} = \sqrt{.5/9} = .236$.

Example 5.3.4: Effect size for contrast freshman and senior of the unbalanced design

This example shows how to calculate the effect size for comparing freshman and senior students using the unbalanced design. From the data, we have the weight $w_1 = .1, w_4 = .4$. The between group variance is $\sigma_b^2 = (4-2)^2/(1/.1 + 1/.4) = 0.32$. For the within group variance, $\sigma_w^2 = (9+9+9+9)/4 = 9$. Therefore, the effect size is $f = \sqrt{\sigma_b^2/\sigma_w^2} = \sqrt{.32/9} = .189$.

Example 5.3.5: Effect size from empirical data

The effect size can also be determined from a set of empirical data. The data could come from a pilot study. From the empirical data, the groups means and variances can be easily obtained. In this example, the data set at http://psychstat.org/anovadata includes two variables - an outcome variable y and a grouping variable group. From the data set, one can calculate the between-group variance (14.58) and within-group variance (4.70), from which the effect size can be obtained as 1.76.

5.3.3 *Effect size calculator*

The specification of the effect size can be assisted by an online calculator. In the interface in Figure 5.1.1, clicking the link "Calculator" brings up the effect size calculator. The calculator allows the calculation of effect sizes using three methods as shown in Figure 5.3.1.

http://psychstat.org/anovaeffect

Method 1 determines an effect size based on the input of the between-group variance and the within-group variance. With the two values, the effect size is calculated and shown after clicking on the "**Calculate**" button.

Method 2 allows a user to input group sample size[4], mean, and variance to calculate both the overall effect size and the effect size for each contrast. By default, one can input data for three groups. However, one can specify any number of groups by inputting it in the *Number of groups* field and clicking the "**Update**" button to show more groups. For example, to obtain the effect sizes based on the unbalanced design in Table 5.3.1, one can input the data as in Figure 5.3.1. At the bottom, the effect sizes for both overall effect and the contrast effects are presented after clicking on the "**Calculate**" button.

[4] Note that both balanced and unbalanced sample sizes can be used.

Effect Size Calculator for One-way ANOVA
Method 2: Use group mean information

Number of groups: 4   Update

Group	Sample size	Mean	Variance
1	10	2	9
2	20	3	9
3	30	3.6	9
4	40	4	9

Calculate

Effect size output

The overall effect size f = **0.2048**
The effect size for Group 1 vs Group 2 is f = **0.0861**
The effect size for Group 1 vs Group 3 is f = **0.1461**
The effect size for Group 1 vs Group 4 is f = **0.1886**
The effect size for Group 2 vs Group 3 is f = **0.0693**
The effect size for Group 2 vs Group 4 is f = **0.1217**
The effect size for Group 3 vs Group 4 is f = **0.0552**

Figure 5.3.1: Effect size calculation based on the input of means

Method 3 allows a user to upload a set of data and calculates effect sizes based on the data directly. Figure 5.3.2 shows the use of the data in Example 5.3.5 and the output of the effect sizes. Note that only registered users can use this method to protect data privacy. The data file has to be in text format where the first column of the data is the outcome variable and the second is the grouping variable. The first line of the data should be the variable names.

The output of Method 3 includes the overall effect size and the effect sizes for all the contrasts. In addition, this method also conducts ANOVA based on the data uploaded and presents the ANOVA output as shown in Figure 5.3.2.

Example data:

```
     y group
22.48831    1
15.48998    1
18.97749    1
16.32764    1
22.64907    1
19.14727    1
```

Effect Size Calculator for One-way ANOVA

Method 3: From empirical data analysis

Upload data file:

Choose File | anovadata1.txt   Calculate

Figure 5.3.2: Effect size calculation based on an empirical data set

Effect size output

The overall effect size f = **1.7613**
The effect size for Group 1 vs 2 is f= **0.3269**
The effect size for Group 1 vs 3 is f= **0.7833**
The effect size for Group 1 vs 4 is f= **1.6567**
The effect size for Group 2 vs 3 is f= **0.4564**
The effect size for Group 2 vs 4 is f= **1.3298**
The effect size for Group 3 vs 4 is f= **0.8734**

Output from ANOVA

```
          Df Sum Sq Mean Sq F value Pr(>F)
group      3 1166.2   388.7   78.59 <2e-16 ***
Residuals 76  375.9     4.9
---
Signif. codes:  0 '***' 0.001 '**' 0.01 '*' 0.05 '.' 0.1 ' ' 1
```

Note that the methods for effect size calculation literally assume that the population effect sizes are the same as the obtained sample effect sizes, which might not be correct.

5.4   Technical Details

Suppose there exists a factor A with k levels or groups. The sample size for each group is $n_g, g = 1, \ldots k$. The total sample size is $n = \sum_{g=1}^{k} n_g$. Let y_{ig} denotes the datum for the ith individual in the gth group. Assume that $y_{ig} \sim N(\mu_g, \sigma^2)$ where μ_g is the group mean of the gth group. ANOVA usually concerns the overall test of equality of the means across groups with

$$H_0 : \mu_1 = \mu_2 = \ldots = \mu_k = \mu$$

indicating that all the group means are the same vs.

$$H_1 : \text{ for at least a pair of } j \text{ and } l; \mu_j \neq \mu_l,$$

or there exist at least two groups with different means.

One can also conduct a test on any contrast

$$H_0 : \sum_{g=1}^{k} c_g \mu_g = c_0$$

where c_0 is often 0 and $\sum_{g=1}^{k} c_g = 0$. The alternative hypothesis can be either two-sided or one-sided

$$H_{11} : \sum_{g=1}^{k} c_g \mu_g \neq c_0$$

or

$$H_{12} : \sum_{g=1}^{k} c_g \mu_g > c_0$$

or

$$H_{13} : \sum_{g=1}^{k} c_g \mu_g < c_0.$$

The hypotheses can be tested by framing the ANOVA as a linear model

$$\mathbf{y} = \mathbf{X}\boldsymbol{\beta} + \boldsymbol{\epsilon}$$

where $\mathbf{y} = (y_{ig})$ is a vector containing the outcome from all k groups and $\mathbf{X}$ is a design matrix. For one-way ANOVA, $\mathbf{X}$ is a matrix of 0 and 1. Specifically,

$$\mathbf{X} = \begin{pmatrix} \mathbf{1}_1 & 0 & \cdots & 0 \\ 0 & \mathbf{1}_2 & \cdots & 0 \\ 0 & 0 & \ddots & 0 \\ 0 & 0 & \cdots & \mathbf{1}_k \end{pmatrix}$$

with $\mathbf{1}_g$ represents a $n_g \times 1$ column vector of ones. The regression coefficient $\boldsymbol{\beta}$ is a vector of group means μ_g. Estimating the regression model gives

$$\begin{aligned} \mathbf{b} &= \hat{\boldsymbol{\beta}} = (\mathbf{X}'\mathbf{X})^{-1}\mathbf{X}'\mathbf{y} \\ \hat{\sigma}^2 &= (\mathbf{y} - \mathbf{X}\mathbf{b})'(\mathbf{y} - \mathbf{X}\mathbf{b})/(n-k). \end{aligned}$$

A general hypothesis can be formed by

$$H_0 : \mathbf{C}\boldsymbol{\beta} = \mathbf{c}_0$$

where $\mathbf{C}$ is a $K \times k$ contrast matrix with $rank(\mathbf{C}) = K \leq k$. Given the normality assumption, the statistic

$$F = \frac{(\mathbf{C}\mathbf{b} - \mathbf{c}_0)'[\mathbf{C}(\mathbf{X}'\mathbf{X})^{-1}\mathbf{C}']^{-1}(\mathbf{C}\mathbf{b} - \mathbf{c}_0)}{(G-1)\hat{\sigma}^2} \tag{5.4.1}$$

follows an F distribution with degrees of freedom K and $n - r$ where r is the rank of $\mathbf{X}$. If the F statistic is larger than the critical value $F_{K, n-r, 1-\alpha}$, one would reject the null hypothesis H_0. For one-way ANOVA, $K = k - 1$ and $r = k$.

Under the alternative hypothesis, the statistic F in Equation 5.4.1 follows a non-central F distribution with the non-central parameter

$$\lambda = (\mathbf{Cb} - \mathbf{c}_0)'[\mathbf{C}(\mathbf{X}'\mathbf{X})^{-1}\mathbf{C}']^{-1}(\mathbf{Cb} - \mathbf{c}_0)/\sigma^2 = nf^2,$$

with f denoting the effect size. Power can be readily calculated using the non-central F distribution. For one-way ANOVA, we have

$$\pi = 1 - F_{k-1,n-k,\lambda}(F_{k-1,n-k,1-\alpha}^{-1}),$$

where F is the cumulative distribution function, $F_{k-1,n-k,1-\alpha}^{-1}$ is the critical value of an F distribution given the probability $1 - \alpha$, and F gives the probability for a given quantile.

Similarly, power for a given contrast can be obtained. For the contrast, power can be calculated based on either a t test or an F test. Depending on the alternative hypothesis, the power is

$$\pi = \begin{cases} 1 - F_{1,n-k,\kappa^2}(F_{1,n-k,1-\alpha}^{-1}) & H_{11} \\ 1 - t_{n-k,\kappa}(t_{n-k,1-\alpha}^{-1}) & H_{12} \\ t_{n-k,\kappa}(t_{n-k,\alpha}^{-1}) & H_{13} \end{cases}$$

where t is the cumulative distribution function for a t distribution. The non-centrality parameter κ for the t distribution and κ^2 for the F distribution are calculated as

$$\kappa = \sqrt{n}(\mathbf{Cb} - \mathbf{c}_0)'\sqrt{[\mathbf{C}(\mathbf{X}'\mathbf{X})^{-1}\mathbf{C}']^{-1}/\sigma_e^2} = \sqrt{n}f,$$

with f denoting the effect size.

5.5   Exercises

1. A researcher plans to design a study with 4 groups: a memory intervention, a reasoning intervention, a processing speed intervention, and a control group. To decide the sample size, he has found the following data in the literature. What would be the total sample size needed for a balanced design to get a power 0.8 at the alpha level 0.05 if assuming the population effect is the same as the sample effect?

	Memory	Reasoning	Speed	Control
Mean	26	25.8	26.2	25.6
s.d.	5.5	5.4	5.3	5.7

Using the same information, what would be the required sample sizes when the alpha level is set at 0.1 and 0.01, respectively?

2. Using the same information in Exercise 1, generate a power curve with the total sample size ranging from 100 to 2000 with an interval of 100. From the power curve, approximately how large is a sample size needed to get a power 0.9?

3. If the researcher is especially interested in the difference between the processing speed group and the control group, what would be the sample size required to detect the speed group has a higher score than the control group?

4. If a researcher can collect more data from the control group with the sample size ratio 1:1:1:3 for the four groups, what would be the power for a total of 1,800 participants?

5. If a researcher can collect more data from the control group with the sample size ratio 1:1:1:3 for the four groups, to get a power 0.8, how many participants are needed?

6   *Statistical Power Analysis for Two-Way and Three-Way ANOVA*

Zhiyong Zhang
Department of Psychology
University of Notre Dame

Two-way analysis of variance (two-way ANOVA) is a generalization of one-way ANOVA where two main effects and their interaction effect can be studied. More general k-way ANOVA can be defined in the same way. The power analysis for ANOVA with more than one factor can be similarly conducted as for the one-way ANOVA. In this chapter, we show how to conduct power analysis for main effects as well as interaction effects in two-way and three-way ANOVA.

6.1   *How to Conduct Power Analysis for Two-way and Three-way ANOVA*

The primary software interface for power analysis for two-way, three-way and more general k-way ANOVA is shown in Figure 6.1.1.. Within the interface, a user can supply different parameter values and select different options for power analysis. Among the five parameters, *Number of groups*, *Total sample size*, *Effect size*, *Significance level*, and *Power*, one and only one can be left blank. To illustrate the meaning of the parameter, consider a three-way ANOVA with three factors A, B, and C. The number of levels (categories or groups) for the three factors is J, K, and L, respectively. As an example, let $J = 3, K = 2$, and $L = 4$.

http://psychstat.org/kanova

- The *Number of groups* is the total number of groups in the design calculated by $J \times K \times L$. For the three-way ANOVA example, the total number of groups is $3 \times 2 \times 4 = 24$. For two-way ANOVA with the first two factors only, the number of groups is $3 \times 2 = 6$.

- The *Sample size* is the **total** number of participants from all groups. The power calculation assumes the equal sample size for all groups. The total sample size is the product of the number of groups and the

Two-Way, three-Way, and k-Way ANOVA

Figure 6.1.1: Software interface of power analysis for two-way and three-way ANOVA

Parameters (Help)

Number of groups	4
Total sample size	100
Numerator df	2
Effect size (f) (Calculator)	0.5
Significance level	0.05
Power	
Power curve	No power curve ▾
Note	Power analysis for k-way A

Calculate

sample size for each group. For example, if 5 subjects are in each of the 24 ($= 3 \times 2 \times 4$) groups, then the total sample size would be $5 \times 24 = 120$.

Multiple sample sizes can be provided in two ways to calculate power for each sample size. First, multiple sample sizes can be supplied and separated by white spaces, e.g., 100 150 200 will calculate power for the three sample sizes 100, 150 and 200. Second, a sequence of sample sizes can be generated using the method s:e:i with s denoting the starting sample size, e as the ending sample size, and i as the interval. Note the values are separated by colon ":". For example, 100:150:10 will generate a sequence of sample sizes - 100 110 120 130 140 150. The default sample size, as shown in Figure 6.1.1, is 100.

- The power is calculated based on an F distribution which requires the numerator and denominator degrees of freedom (df). The *Numerator df* depends on the effect to be analyzed and needs to be provided. For the main effect, it is the number of levels - 1. For example, if power is calculated for the main effect of A, then the numerator df is $J - 1 = 3 - 1 = 2$. For B and C, the dfs are 1 and 3, respectively. For the interaction effect, the numerator df is calculated as $(J - 1) \times (K - 1) \times (L - 1)$ for the three-way interaction. For two-way interaction, it is calculated the same way. For example, for the interaction between A and B, the numerator df is $(J - 1) \times (K - 1)$. Using the example, the numerator df for the three-way interaction of A, B, and C is $(3 - 1) \times (2 - 1) \times (4 - 1) = 6$. For the two-way interaction A by B, B by C, and A by C, the numerator dfs are 2, 3, and 6, respectively.

- The *Effect size*[1] specifies the population group difference. Multiple effect sizes or a sequence of effect sizes can also be supplied using the same format as for sample size. By default, the value is 0.5. Determining the effect size is critical but not trivial work. To help a user to obtain effect sizes, a calculator has been developed for two-way ANOVA and can be used by clicking the link "**Calculator**".

- The *Significance level*[2] for power calculation is needed but usually set at the default 0.05.

[2] Type I error rate or alpha level

- The *Power* specifies the desired statistical power.

- In addition to the required input, one can also request the plot of a power curve if multiple sample sizes or effect sizes are provided.

- A note (less than 200 characters) can be provided to provide basic information on the analysis for future reference for registered users.

6.1.1 *Examples*

A researcher is interested in understanding whether education and gender are related to the amount of technical knowledge. In order to find it out, he plans to collect data from three education level groups and two gender groups. For each combination of education and gender groups, he plans to recruit 20 students. Based on the literature, he knows the effect size for the education factor is about 0.2. What would be his power to find the significant education effect?

Example 6.1.1: Power for a main effect of two-way ANOVA

The input and output for calculating power for this study are given in Figure 6.1.2. In the field of *Number of groups*, input 6 ($= 3 \times 2$), the total number of groups; in the field of *Sample size*, input 120 ($= 20 \times 6$), the total sample size of the 6 groups; in the field of *Numerator df*, input 2 ($= 3 - 1$), the total number of education groups minus 1; and in the field of *Effect size*, input 0.2, the expected effect size. The default significance level 0.05 is used although one can change it to a different value. The field for *Power* is left blank because it will be calculated. A simple note "2-way ANOVA main effect" is also added in the *Note* field. By clicking the "**Calculate**" button, the statistical power is given in the output immediately. For the current design, the power is 0.4758. Note that in the output, the denominator degrees of freedom (ddf) were calculated and provided in the output.

The researcher in Example 6.1.1 is also interested in the power to detect the interaction effect with the effect size 0.4. The input and output for calculating power for the interaction are given in Figure 6.1.3. Note that the *Sample size* and the *Number of groups* are the same. For

Example 6.1.2: Power for the interaction effect of two-way ANOVA

Two-Way, three-Way, and k-Way ANOVA

Figure 6.1.2: Input and output for calculating power for two-way ANOVA main effect in Example 6.1.1

Parameters (Help)

Number of groups	6
Total sample size	120
Numerator df	2
Effect size (f) (Calculator)	0.2
Significance level	0.05
Power	
Power curve	No power curve ▼
Note	2-way ANOVA main effect

Calculate

Output

```
Multiple way ANOVA analysis

      n ndf ddf    f ng alpha  power
    120   2 114 0.2  6  0.05 0.4758


NOTE: Sample size is the total sample size
URL: http://psychstat.org/kanova
```

the *Numerator df*, it is also 2 but calculated as $(3-1) \times (2-1) = 2$. The power for the interaction effect is 0.9789 as shown in 6.1.3.

Example 6.1.3: Power for three-way interaction in three-way ANOVA

Another researcher believes that in addition to education and gender, the geographical location is also related to technical knowledge. Therefore, she plans to expand the study in Example 6.1.1 to collect data from three locations: Eastern, Western and Central US. She is particularly interested in the interaction among education, gender and geographical location. If she found that the effect size for the three-way interaction would be 0.3 based on the existing literature. What would be her power if she keeps collecting data from 20 participants in each group?

The input and output for this analysis are given in Figure 6.1.4. In the field of *Number of groups*, input 18 $(= 3 \times 2 \times 3)$, the total number of groups; in the field of *Sample size*, input 360 $(= 20 \times 18)$, the total sample size of the 18 groups; in the field of *Numerator df*, input 4 $[= (3-1) \times (2-1) \times (3-1)]$; and in the field of *Effect size*, input 0.3.

Two-Way, three-Way, and k-Way ANOVA

Parameters (Help)

Number of groups	6
Total sample size	120
Numerator df	2
Effect size (f) (Calculator)	0.4
Significance level	0.05
Power	
Power curve	No power curve ▾
Note	2-way ANOVA interaction e

Calculate

Output

```
Multiple way ANOVA analysis

     n ndf ddf    f ng alpha  power
   120   2 114 0.4  6  0.05 0.9789


NOTE: Sample size is the total sample size
URL: http://psychstat.org/kanova
```

Based on the input, the obtained power is 0.9983.

As in one-way ANOVA, we can also obtain a power curve and estimate the minimum detectable effect size. However, power analysis for contrasts is not currently available in WebPower.

6.2   *Using R for Power Analysis for Two-way and Three-way ANOVA*

The online power analysis is carried out using the R package WebPower on a Web server. The package can be directly used within R for power analysis for k-way ANOVA. Specifically, the function wp.kanova is used. The basic usage of the function is provided below.

```
wp.kanova(n = NULL, ndf = NULL, f = NULL, ng = NULL, alpha =
    0.05,     power = NULL)
```

For example, the R input and output for the examples discussed in the previous section are given below.

n: sample size
ndf: numerator degrees of freedom
 f: effect size
ng: number of groups
alpha: significance level
power: statistical power

Two-Way, three-Way, and k-Way ANOVA

Figure 6.1.4: Input and output for calculating power of three-way interaction in ANOVA in Example 6.1.3

Parameters (Help)

Number of groups	18
Total sample size	360
Numerator df	4
Effect size (f) (Calculator)	0.3
Significance level	0.05
Power	
Power curve	No power curve ▾
Note	3-way ANOVA interaction

Calculate

Output

```
Multiple way ANOVA analysis

      n ndf ddf   f ng alpha  power
    360   4 342 0.3 18  0.05 0.9983


NOTE: Sample size is the total sample size
URL: http://psychstat.org/kanova
```

```
> ## Main effect of two-way ANOVA
> wp.kanova(n=120, ndf=2, f=0.2, alph=0.05, ng=6)
Multiple way ANOVA analysis

      n ndf ddf   f ng alpha  power
    120   2 114 0.2  6  0.05 0.4758

NOTE: Sample size is the total sample size
URL: http://psychstat.org/kanova
>
> ## Interaction effect of two-way ANOVA
> wp.kanova(n=120, ndf=2, f=0.4, alph=0.05, ng=6)
Multiple way ANOVA analysis

      n ndf ddf   f ng alpha  power
    120   2 114 0.4  6  0.05 0.9789

NOTE: Sample size is the total sample size
URL: http://psychstat.org/kanova
```

```
>
>
> ## Interaction effect of three-way ANOVA
> wp.kanova(n=360, ndf=4, f=0.3, alph=0.05, ng=18)
Multiple way ANOVA analysis

      n ndf ddf   f ng alpha  power
    360   4 342 0.3 18  0.05 0.9983

NOTE: Sample size is the total sample size
URL: http://psychstat.org/kanova
```

6.3   Effect Size for Two-way and Three-way ANOVA

We use the statistic f as the measure of effect size for two-way and more generally k-way ANOVA as in Cohen (1988, p. 275). The f is the ratio between the standard deviation of the effect to be tested σ_b (or the standard deviation of the group means, or between-group standard deviation) and the common standard deviation within the populations (or the standard deviation within each group, or within-group standard deviation) σ_w such that

$$f = \frac{\sigma_b}{\sigma_w}.$$

Given the two quantities σ_b and σ_w, the effect size can be determined. Cohen defined the size of effect as: small 0.1, medium 0.25, and large 0.4. We show how to calculate effect sizes for two-way and three-way ANOVA through examples.

6.3.1   Effect size for two-way ANOVA

To show how to calculate the effect sizes, more precisely sample effect sizes, under different conditions, consider the example on the relationship between education and geographical location and the amount of technical knowledge. Suppose there are three education levels: below high school (<HS), high school (HS), and above high school (>HS) and there are also three geographical locations: Eastern, Western, and Central United States. Furthermore, for each combination of levels for the two factors, there are data from 5 participants as shown in Table 6.3.1.

Based on the raw data, we can calculate different quantities related to the calculation of effect size. The algebraic expression for each quantity is given in Table 6.3.2 (using the notation in Maxwell & Delaney, 2003). If we know these quantities, there is no need for raw data for the calculation of effect size. If they are the population values, population effect sizes are obtained. Otherwise, sample effect sizes are calculated.

We intentionally ignore the difference between population and sample effect sizes here.

	Eastern			Western			Central		
	<HS	HS	>HS	<HS	HS	>HS	<HS	HS	>HS
	14	15	12	10	12	10	8	9	6
	17	11	10	16	14	3	10	6	10
	10	12	10	19	10	6	12	7	8
	13	10	9	20	11	8	14	12	9
	12	9	11	19	13	2	11	11	7

Table 6.3.1: Example data for effect size calculation for two-way ANOVA.

Quantity		Expression
μ_{jk}	$=$	$\frac{1}{n_{jk}} \sum_{i=1}^{n_{jk}} y_{ijk}$
$\mu_{..}$	$=$	$\frac{1}{JK} \sum_{j=1}^{J} \sum_{k=1}^{K} \mu_{jk}$
$\mu_{j.}$	$=$	$\frac{1}{K} \sum_{k=1}^{K} \mu_{jk}$
$\mu_{.k}$	$=$	$\frac{1}{J} \sum_{j=1}^{J} \mu_{jk}$
α_j	$=$	$\mu_{j.} - \mu_{..}$
β_k	$=$	$\mu_{k.} - \mu_{..}$
$(\alpha\beta)_{jk}$	$=$	$\mu_{jk} - (\mu_{..} + \alpha_j + \beta_k)$

Table 6.3.2: Algebraic expression for some quantities used in two-way ANOVA

With real data, the notations have to be replaced by the sample ones. Based on the data in Table 6.3.1, we get the following information in Table 6.3.3.

	Eastern	Western	Central	Average ($\mu_{j.}$)	α_j
<HS	13.2(2.6)	11.4(2.3)	10.4(1.1)	11.67	0.82
HS	16.8(4.1)	12(1.6)	5.8(3.3)	11.53	0.69
<HS	11(2.2)	9(2.5)	8(1.6)	9.33	-1.51
$\mu_{.k}$	13.67	10.8	8.07	10.84	
β_k	2.82	-0.04	-2.78		

Table 6.3.3: Summary data for effect size calculation. Numbers in the parentheses are standard deviations.

To get the effect size, we first need to calculate σ_w. It can be calculated as the square root of the average of the variances of all groups. In this case, it would be

$$\sigma_w = \sqrt{\frac{\sum s_{jk}^2}{JK}} = \sqrt{\frac{2.6^2 + 2.3^2 + \ldots + 1.6^2}{9}} = 2.53$$

Now we move on to the calculation of σ_b. Suppose we are interested in the **main effect of geographical location**. Then, we need to calculate the variance for the factor of geographical location using the marginal means. That is

$$\sigma_b = \sqrt{\frac{\sum \beta_k^2}{3}} = \sqrt{\frac{1}{3}[(13.67 - 10.84)^2 + (10.8 - 10.84)^2 + (8.07 - 10.84)^2]} = 2.29$$

Together, this will give the effect size

$$f = \frac{\sigma_b}{\sigma_w} = \frac{2.29}{2.53} = 0.9.$$

For the **interaction effect**, we calculate

$$(ab)_{jk} = \mu_{jk} - (\mu_{..} + \alpha_j + \beta_k)$$

and then the standard deviation as

$$\sigma_b = \sqrt{\frac{\sum (ab)_{jk}^2}{JK}}.$$

In this example, $\sigma_b = 1.58$. Thus, the effect size is $f = 1.58/2.53 = .62$.

6.3.2 *Effect size calculator*

The specification of the effect size can be assisted by an online calculator. In the interface in Figure 6.1.1, clicking the link "Calculator" brings up the effect size calculator. The calculator allows the calculation of effect sizes using three methods as shown in Figure 6.3.1. The first and second methods assume a balanced design.

http://psychstat.org/anovaeffect

Method 1 allows a user to input marginal means and common variance (σ_w^2) to calculate the effect size of the two main effects. By default, one can input data for three groups for each factor. However, one can specify any number of groups for the two factors by inputting them and clicking the "**Update**" button.

For example, to obtain the effect sizes for the data given in Table 6.3.3, the input for *Number of groups for factor A* and *Number of groups for factor B* are both 3. The inputs for *Factor A* are the marginal means, 11.67, 11.53 and 9.33, the input for *Factor B* are 13.67, 10.8 and 8.07. The inputs for *Error variance* is $2.53^2 = 6.4$. By clicking the "**Calculate**" button, the effect sizes are presented at the bottom as in Figure 6.3.2. In the output, the effect size for factor A is $f = 0.4236$, and the effect size for factor B is $f = 0.9038$.

Method 2 allows a user to input cell means and common variance (σ_w^2) to calculate the effect sizes for Factor A, Factor B and the interaction between A and B. Similarly, the default number of groups for both factors are 3 but one can specify any number of groups by changing the inputs and clicking the "**Update**" button.

For example, to obtain the effect sizes for the data given in Table 6.3.3, the input for *Number of groups for factor A* and *Number of groups for factor B* are both 3. The input for *A1* line is 13.2, 11.4 and 10.4, the input for *A2* line is 16.8, 12 and 5.8, the input for *A3* line is 11, 9 and 8. The input for *Common variance* is 6.4. By clicking the "**Calculate**" button, the effect sizes are presented at the bottom as in Figure 6.3.3. In the output, the effect size for factor A is $f = 0.4229$, the effect size for factor B is $f = 0.9038$, and the effect size for the interaction term is $f = 0.6246$.

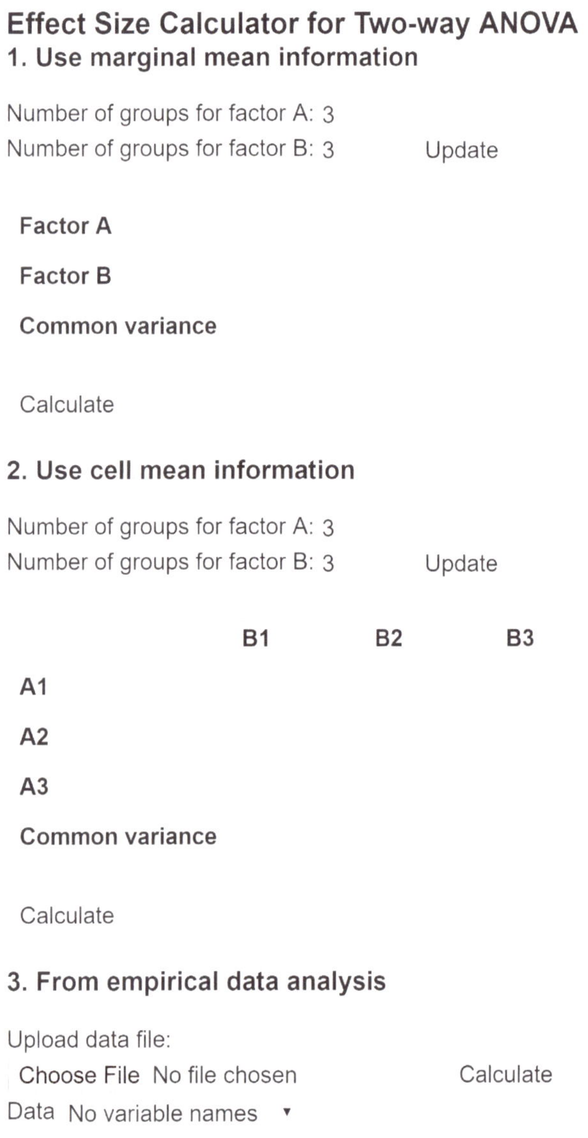

Effect Size Calculator for Two-way ANOVA
1. Use marginal mean information

Number of groups for factor A: 3
Number of groups for factor B: 3   Update

Factor A

Factor B

Common variance

Calculate

2. Use cell mean information

Number of groups for factor A: 3
Number of groups for factor B: 3   Update

	B1	B2	B3
A1			
A2			
A3			
Common variance			

Calculate

3. From empirical data analysis

Upload data file:
  Choose File No file chosen   Calculate
Data No variable names ▾

Figure 6.3.1: The interface for effect size calculation for two-way ANOVA

Effect Size Calculator for Two-way ANOVA
1. Use marginal mean information

Figure 6.3.2: Effect size calculation based
on the input of marginal means

Number of groups for factor A: 3
Number of groups for factor B: 3   Update

Factor A	11.67	11.53	9.33
Factor B	13.67	10.8	8.07
Common variance	6.4		

Calculate

Effect size output

The total number of groups = 9
For Factor A
The effect size for Factor A: f = **0.4236**
The numerator df = 2
For Factor B
The effect size for Factor B: f = **0.9038**
The numerator df = 2

Effect Size Calculator for Two-way ANOVA
2. Use cell mean information

Number of groups for factor A: 3
Number of groups for factor B: 3   Update

	B1	B2	B3
A1	13.2	11.4	10.4
A2	16.8	12	5.8
A3	11	9	8
Common variance	6.4		

Calculate

Effect size output

The total number of groups = 9
<u>For Factor A</u>
The effect size for Factor A: f = **0.4229**
The numerator df = 2
<u>For Factor B</u>
The effect size for Factor B: f = **0.9038**
The numerator df = 2
<u>For Interaction A*B</u>
The effect size for A*B: f = **0.6246**
The numerator df = 4

Figure 6.3.3: Effect size calculation based on the input of cell means

Method 3 allows a user to upload a set of data and calculates effect sizes from data directly. Figure 8.3.4 shows the use of the data in Table 8.3.2 and the output of the effect sizes. Note that only registered users can use this method to protect data privacy. In addition to the effect sizes, this method also conducts a two-way ANOVA analysis based on the data uploaded.

The data file has to be in plain text format where the first column of the data is the outcome variable, and the second and third columns are the two factors. If the first line of the data contains variable names, then one can choose "With variable names" from the drop-down menu of the Data field. In the output, one can obtain the estimated effect sizes and the numerator degrees of freedom.

Example data:

y	group
22.48831	1
15.48998	1
18.97749	1
16.32764	1
22.64907	1
19.14727	1

One such data set can be seen at http://psychstat.org/2anovadata.

Effect Size Calculator for Two-way ANOVA

3. From empirical data analysis

Upload data file:
Choose File twowayanovadata.txt   Calculate
Data No variable names ▾

Effect size output

The total number of groups = **9**
For Factor A
The effect size for Factor A: f = **0.903**
The numerator df = **2**
For Factor B
The effect size for Factor B: f = **0.4225**
The numerator df = **2**
For Interaction A*B
The effect size for A*B: f = **0.6241**
The numerator df = **4**

Output from ANOVA

```
          Df Sum Sq Mean Sq F value   Pr(>F)
g1         2 235.24  117.62  18.347 3.21e-06 **
g2         2  51.51   25.76   4.017  0.02662 *
g1:g2      4 112.36   28.09   4.381  0.00547 **
Residuals 36 230.80    6.41
---
Signif. codes:  0 '***' 0.001 '**' 0.01 '*' 0.05
```

6.3.3 Effect size for three-way ANOVA

We now illustrate how to calculate the main and interaction effect sizes for a three-way ANOVA. Suppose we have the data in Table 6.3.4. Note that in addition to the gender and education factors, we consider a third factor – geographical location as in Example 6.3.4.

	Eastern			Western			Central		
	<HS	HS	>HS	<HS	HS	>HS	<HS	HS	>HS
Male	14	15	12	10	12	10	8	9	6
	17	11	10	16	14	3	10	6	10
	10	12	10	19	10	6	12	7	8
	13	10	9	20	11	8	14	12	9
	12	9	11	19	13	2	11	11	7
Female	17	16	14	11	14	11	11	10	9
	18	14	13	16	14	5	13	9	13
	13	15	13	20	11	8	12	8	10
	13	10	9	20	11	11	15	15	9
	13	12	13	20	16	4	11	13	7

Table 6.3.4: Example data for effect size calculation for a three-way ANOVA. The data are measures of technical knowledge.

To calculate the effect sizes, we need to calculate some summary statistics as in the two-way ANOVA analysis. The expressions for the statistics are given in Table 6.3.5.

Quantity		Expression
μ_{jkl}	=	$\sum_{i=1}^{n_{jkl}} y_{ijkl} / n_{jkl}$
$\mu_{...}$	=	$\sum_{j=1}^{J} \sum_{k=1}^{K} \sum_{l=1}^{L} \mu_{jkl} / (JKL)$
$\mu_{j..}$	=	$\sum_{k=1}^{K} \sum_{l=1}^{L} \mu_{jkl} / (KL)$
$\mu_{.k.}$	=	$\sum_{j=1}^{J} \sum_{l=1}^{L} \mu_{jkl} / (JL)$
$\mu_{..l}$	=	$\sum_{j=1}^{J} \sum_{k=1}^{K} \mu_{jkl} / (JK)$
$\mu_{jk.}$	=	$\sum_{l=1}^{L} \mu_{jkl} / L$
$\mu_{.kl}$	=	$\sum_{j=1}^{J} \mu_{jkl} / J$
$\mu_{j.l}$	=	$\sum_{k=1}^{K} \mu_{jkl} / K$
α_j	=	$\mu_{j..} - \mu_{...}$
β_k	=	$\mu_{.k.} - \mu_{...}$
γ_l	=	$\mu_{..l} - \mu_{...}$
$(\alpha\beta)_{jk}$	=	$\mu_{jk.} - (\mu_{...} + \alpha_j + \beta_k)$
$(\alpha\gamma)_{jl}$	=	$\mu_{j.l} - (\mu_{...} + \alpha_j + \gamma_l)$
$(\beta\gamma)_{kl}$	=	$\mu_{.kl} - (\mu_{...} + \beta_k + \gamma_l)$
$(\alpha\beta\gamma)_{jkl}$	=	$\mu_{jkl} - [\mu_{...} + \alpha_j + \beta_k + \gamma_l + (\alpha\beta)_{jk} + (\alpha\gamma)_{jl} + (\beta\gamma)_{kl}]$

Table 6.3.5: Algebraic expression for statistics used in three-way ANOVA

Based on the example data, the different statistics can be calculated as shown in Table 6.3.6. Again, we do not distinguish population and sample here. For power analysis, population values should be assumed although they can be hypothesized to equal the estimated value.

Main effect of education

Suppose we are interested in the effect size of the main effect of education. First, we need to get the within group or common variance

Table 6.3.6: Summary statistics. Gender (j), Location (k), Education (l)

		Eastern <HS	Eastern HS	Eastern >HS	Western <HS	Western HS	Western >HS	Central <HS	Central HS	Central >HS
μ_{jkl}	Male	13.2	11.4	10.4	16.8	12	5.8	11	9	8
	Female	14.8	13.4	12.4	17.4	13.2	7.8	12.4	11	9.6
σ^2_{jkl}	Male	5.4	4.2	1.0	13.4	2.0	9.0	4.0	5.2	2.0
	Female	5.0	4.6	3.0	12.6	3.8	8.6	2.2	6.8	3.8
$\mu_{.kl}$		14	12.4	11.4	17.1	12.6	6.8	11.7	10	5.0
$\mu_{j.l}$		13.7	10.8	8.1	14.9	12.5	9.9			
$\mu_{jk.}$		11.7	11.5	9.3	13.5	12.8	11.0			
$\mu_{..l}$		14.3	11.7	9.0						
$\mu_{.k.}$		12.6	12.2	10.2						
$\mu_{j..}$		10.8	12.4							
$\mu_{...}$		11.6								
γ_l		2.6	0.0	-2.6						
β_k		1.0	0.5	-1.5						
α_j		-0.8	0.8	1.4						
$(\alpha\gamma)_{kl}$		-1.2	-0.2	-0.1	2.3	0.4	-2.7	-1.1	-0.2	1.3
$(\beta\gamma)_{jl}$		0.2	-0.1	0.0	-0.2	0.1	0.1			
$(\alpha\beta)_{jk}$		-0.1	0.2	0.07	0.1	-0.2	0.0			
$(\alpha\beta\gamma)_{jkl}$		-0.07	0.00	-0.07	0.13	0.10	-0.23	-0.07	-0.10	0.17
		0.07	0.00	0.07	-0.13	-0.10	0.23	0.07	0.10	-0.17

σ^2_w, which is

$$\sigma^2_w = \frac{1}{JKL} \sum_{j=1}^{J} \sum_{k=1}^{K} \sum_{l=1}^{L} \sigma^2_{jkl} = \frac{1}{18}(5.4 + 4.2 + \ldots + 6.8 + 3.8) = 5.37.$$

For the effect size of the main effect of education, we then need to calculate σ^2_b. This can be calculated using the marginal means for education as

$$\sigma^2_b = \frac{1}{L} \sum_{l=1}^{L} (\mu_{..l} - \mu_{...})^2 = \frac{1}{3}[(14.3 - 11.6)^2 + (11.7 - 11.6)^2 + (9.0 - 11.6)^2] = 4.69.$$

Therefore, the effect size is

$$f = \frac{\sigma_b}{\sigma_w} = \sqrt{\frac{4.69}{5.47}} = 0.9345.$$

Interaction effect between gender and location

We can calculate the interaction between any two factors. For the interaction for gender and location, we first need to calculate $(\alpha\beta)_{jk}$. Then, we have

$$\sigma_b^2 = \frac{1}{JK}\sum_{j=1}^{J}\sum_{k=1}^{K}[(\alpha\beta)_{jk}]^2 = [(-.1)^2+0.2^2+0^2+.1^2+(-.2)^2+0^2]/6 = 0.0156.$$

Therefore, the effect size is

$$f = \frac{\sigma_b}{\sigma_w} = \sqrt{\frac{0.0156}{5.47}} = 0.054.$$

Three-way interaction effect size

To get the three-way interaction effect size, we calculate σ_b^2 as

$$\sigma_b^2 = \frac{1}{JKL}\sum_{j=1}^{J}\sum_{k=1}^{K}\sum_{l=1}^{L}[(\alpha\beta\gamma)_{jkl}]^2 = \frac{1}{18}[(-.07)^2+0^2+\ldots+.1^2+(-.17)^2] = 0.0148.$$

Therefore, the effect size is

$$f = \frac{\sigma_b}{\sigma_w} = \sqrt{\frac{0.0148}{5.47}} = 0.0525.$$

6.4   Technical Details

The power analysis for two-way, three-way ANOVA as well as k-way ANOVA uses the same method as for the one-way ANOVA discussed in section 5.4.

6.5   Exercises

1. Using the information in Table 6.3.3, calculate the power for the two main effects and one interaction effect. Find the sample size to obtain a power 0.8.

2. Using the information in Table 6.3.6, calculate the power for the three main effects and all the interaction effects. Find the sample size to obtain a power 0.8.

7 Statistical Power Analysis for One-way ANOVA with Binary or Count Data

Part of the material in this chapter is from Mai & Zhang (2017).

Yujiao Mai and Zhiyong Zhang
Department of Psychology
University of Notre Dame

Comparison of population means is one of the essential statistical analyses in quantitative research (Moore et al., 2013). For comparing means of three or more groups, analysis of variance (ANOVA) is most frequently used (Howell, 2012). Typically, it is used on continuous data and produces an F-statistic as the ratio of the between-group variance to the within-group variance that follows an F-distribution. To use the F-test for ANOVA, three assumptions must be satisfied. They are independence and normality of observations, and equality of variances across groups. In practice, studies with even continuous data cannot always meet all three assumptions. For binary or count data, the assumption of normality is apparently violated. Therefore, it is improper to use classical ANOVA to compare means of binary or count data. Furthermore, the corresponding power analysis is expected to be problematic. To address this issue, Mai & Zhang (2017) introduced analogous ANOVA tables with a closed-form likelihood ratio test statistics for comparing means with binary and count data, and developed the corresponding power analysis methods. In this chapter, we use the methods developed by Mai & Zhang (2017) to conduct the power analysis for binary and count data.

7.1 How to Conduct Power Analysis for One-way ANOVA with Binary Data

Figure 7.1.1 displays the primary software interface for power analysis for one-way ANOVA with binary data. Within the interface, a user can supply different parameter values and select different options for power analysis. Among the five parameters, *Number of groups*[1], *Sample size*, *Effect size*, *Significance level*, and *Power*, one and only one can be left blank.

One-way ANOVA with Binary Data

Parameters (Help)

Number of groups	4
Sample size	100
Effect size (Calculator)	0.5
Significance level	0.05
Power	
Type of analysis	Overall ↕
Power curve	No power curve ↕
Note	Binary ANOVA

Calculate

Figure 7.1.1: Software interface of power analysis for one-way ANOVA with binary data.

- The *Number of groups* tells how many groups are used in the study design.

- The *Sample size* is the total number of participants from all groups. For example, if the sample size for each group is 25 for 4 groups, the total sample size is 100. Multiple sample sizes can be provided in two ways to calculate power for each sample size. First, multiple sample sizes can be supplied and separated by white spaces, e.g., 100 150 200 will calculate power for the three sample sizes 100, 150, and 200. A sequence of sample sizes can also be generated using the method s:e:i with s denoting the starting sample size, e as the ending sample size, and i as the interval. Note that the values are separated by colons. For example, 100:150:10 will generate a sequence of sample sizes - 100 110 120 130 140 150. The default sample size, as shown in Figure 7.1.1, is 100.

- The *Effect size* specifies the population group difference. Multiple effect sizes or a sequence of effect sizes can also be supplied in the same way as sample size. By default, the value is 0.5. Determining the effect size is critical but not trivial work. To help a user obtain effect sizes, a calculator has been developed and can be used by clicking the link "Calculator".

- The *Significance level* for power calculation is required but usually set at the default 0.05.

- The *Power* specifies the desired statistical power. The software currently can calculate power only for the overall effect.

- In addition to the required input, one can also request the plot of a power curve if multiple sample sizes or effect sizes are provided.

- A note (less than 200 characters) can also be used to provide basic information on the analysis which will be saved for a registered user for future reference.

7.1.1   *Examples*

A student hypothesizes that freshman, sophomore, junior and senior college students have different proportions of passing a reading exam. Based on his prior knowledge, he expects that the effect size is about 0.15. If he plans to check the success-failure records of the reading exam with 25 students in each student group, what is the power for him to find the significant difference among the four groups?

The input and output for calculating the power for this study are given in Figure 7.1.2. In the field of *Number of groups*, input 4, the total number of groups; in the field of *Sample size*, input 100, the total sample size of the four groups; and in the field of *Effect size*, input 0.15, the expected effect size. The default significance level 0.05 is used although one can change it to a different value. The field for *Power* is left blank because it is to be calculated. A simple note "Binary ANOVA" is also added in the *Note* field. By clicking the "Calculate" button, we obtain the power 0.5723.

Example 7.1.1: Calculate power given sample size and effect size

A power curve is a line plot of statistical power along with given sample sizes. In Example 7.1.1, the power is 0.572 with the sample size 25 in each group. What is the power for a different sample size, say, 50 in each group? One can investigate the power of different sample sizes and plot a power curve.

The input and output for calculating power for the study in Example 7.1.1 with a sequence of sample sizes from 100 to 200 (25 to 50 each group) with an interval of 20 are given in Figure 7.1.3. Note that in the *Sample size* field, the input is 100:200:20. The output lists the power for each sample size from 100 to 200 with the interval 20. Especially, with the sample size 180 (45 for each group), the power is about 0.845. As we choose "Show power curve" from the *Power curve* drop-down menu in the input, the power curve is displayed at the bottom of the output as

Example 7.1.2: Power curve

One-way ANOVA with Binary Data

Figure 7.1.2: Input and output for calculating power for one-way ANOVA with binary data in Example 7.1.1.

Parameters (Help)

Number of groups   4

Sample size   100

Effect size (Calculator)   0.15

Significance level   0.05

Power

Type of analysis   Overall ⬍

Power curve   No power curve   ⬍

Note   Binary ANOVA

Calculate

Output

```
One-way Analogous ANOVA with Binary Data

   k    n     V alpha   power
   4  100  0.15   0.05  0.5723


NOTE: n is the total sample size
URL: http://psychstat.org/anovabinary
```

shown in Figure 7.1.4. The power curve can be used for interpolation. For example, to get a power 0.8, about 45 students are needed for each group.

Example 7.1.3: Calculate sample size given power and effect size

In practice, a power 0.8 is often desired. Given the power, the sample size can also be calculated as shown in Figure 7.1.5. In this situation, the *Sample size* field is left blank while in the *Power* field, the value 0.8 is input. In the output, we can see a sample size 162 is needed to obtain a power 0.8, that is about 41 for each group.

Example 7.1.4: Minimum detectable effect size

One can also calculate the minimum effect to achieve a certain power given a sample size. As shown in Figure 7.1.6, we leave the *Effect size* field blank but provide information on *Sample size* (100) and *Power* (0.8). In the output, we can see the obtained effect size is 0.1906. Therefore,

One-way ANOVA with Binary Data

Figure 7.1.3: Input and output for power curve of one-way ANOVA with binary data in Example 7.1.2.

Parameters (Help)

Number of groups	4
Sample size	100:200:20
Effect size (Calculator)	0.15
Significance level	0.05
Power	
Type of analysis	Overall ⬍
Power curve	Show power curve ⬍
Note	Binary ANOVA

Calculate

Output

```
One-way Analogous ANOVA with Binary Data

k   n     V alpha  power
4 100 0.15  0.05 0.5723
4 120 0.15  0.05 0.6602
4 140 0.15  0.05 0.7346
4 160 0.15  0.05 0.7959
4 180 0.15  0.05 0.8452
4 200 0.15  0.05 0.8840


NOTE: n is the total sample size
URL: http://psychstat.org/anovabinary
```

to get a power 0.8 with the sample size 100, the population effect size has to be at least 0.191.

7.2 Using R WebPower for One-way ANOVA with Binary Data

The online power analysis is carried out using the R package WebPower on a Web server. The package can be directly used within R for power analysis for one-way ANOVA with binary data. Specifically, the function wp.anova.binary is used. For example, the R input and output for Example 7.1.2 are given below. When the sample size is provided, the power will be calculated. The basic usage of the function is given below.

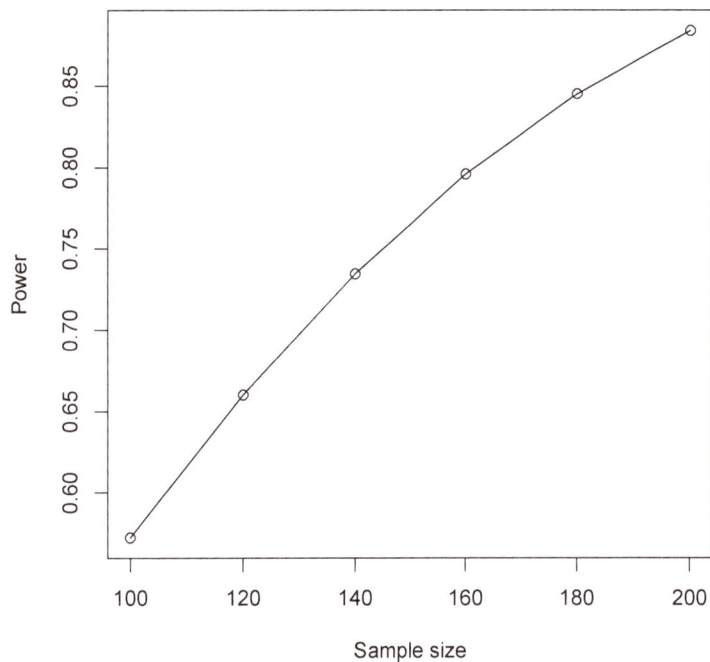

Figure 7.1.4: Power curve for one-way ANOVA with binary data

```
wp.anova.binary(k = NULL, n = NULL, V = NULL, alpha = 0.05, power
    = NULL)
```

The examples in the previous section are included below.

```
> wp.anova.binary(k = 4, n = 100, V = 0.15, alpha = 0.05, power =
    NULL)

Power for One-way ANOVA with Binary Data

      k       n       V     alpha    power
      4      100     0.15    0.05    .5723443

NOTE: n is the total sample size adding all groups (overall)
WebPower URL: http://psychstat.org/anovabinary


>## power curve
> powers <- wp.anova.binary(k = 4, n = seq(100,200,10), V = 0.15,
    alpha = 0.05, power = NULL)
> powers

Power for One-way ANOVA with Binary Data
```

V: effect size
k: number of groups
n: sample size
alpha: significance level
power: statistical power

One-way ANOVA with Binary Data

Figure 7.1.5: Input and output for sample size calculation of one-way ANOVA with binary data in Example 7.1.3.

Parameters (Help)

Number of groups	4
Sample size	
Effect size (Calculator)	0.15
Significance level	0.05
Power	0.8
Type of analysis	Overall ⬍
Power curve	No power curve ⬍
Note	Binary ANOVA

Calculate

Output

```
One-way Analogous ANOVA with Binary Data

    k     n     V alpha power
    4 161.5 0.15  0.05   0.8


NOTE: n is the total sample size
URL: http://psychstat.org/anovabinary
```

```
      k       n       V       alpha     power
      4       100     0.15    0.05      0.5723443
      4       110     0.15    0.05      0.6179014
      4       120     0.15    0.05      0.6601594
      4       130     0.15    0.05      0.6990429
      4       140     0.15    0.05      0.7345606
      4       150     0.15    0.05      0.7667880
      4       160     0.15    0.05      0.7958511
      4       170     0.15    0.05      0.8219126
      4       180     0.15    0.05      0.8451603
      4       190     0.15    0.05      0.8657970
      4       200     0.15    0.05      0.8840327


NOTE: n is the total sample size adding all groups (overall)
WebPower URL: http://psychstat.org/anovabinary

> plot(powers)  ## generate power curve
```

Furthermore, we can also use R to estimate either the required sample size or the minimum detectable effect size.

One-way ANOVA with Binary Data

Parameters (Help)

Number of groups	4
Sample size	100

Effect size (Calculator)

Significance level	0.05
Power	0.8
Type of analysis	Overall ⬍
Power curve	No power curve ⬍
Note	Binary ANOVA

Calculate

Output

```
One-way Analogous ANOVA with Binary Data

   k   n      V alpha power
   4 100 0.1906  0.05   0.8


NOTE: n is the total sample size
URL: http://psychstat.org/anovabinary
```

Figure 7.1.6: Input and output for effect size of one-way ANOVA with binary data in Example 7.1.4.

```
> wp.anova.binary(k = 4, n = NULL, V = 0.15, alpha = 0.05, power
    = 0.8)

Power for One-way ANOVA with Binary Data

     k      n      V      alpha  power
     4      161.5195      0.15   0.05   0.8

NOTE: n is the total sample size adding all groups (overall)
WebPower URL: http://psychstat.org/anovabinary


> wp.anova.binary(k = 4, n = 100, V = NULL, alpha = 0.05, power =
    0.8)

Power for One-way ANOVA with Binary Data

     k      n      V      alpha  power
     4      100    0.1906373     0.05   0.8

NOTE: n is the total sample size adding all groups (overall)
```

7.3   Calculate Effect Size for One-way ANOVA with Binary Data

For the purpose of power analysis, in this study, we use a standardized effect-size measure like Cramer's V, which is a member of the r family effect size (Ellis, 2010). It is also an adjusted version of phi coefficient ϕ that is frequently reported as the measure of effect size for a chi-square test (Cohen, 1988; Ellis, 2010; Fleiss, 1994). It can be viewed as the association between two variables as a percentage of the maximum possible variation in one variable accounted for by the other variable. In the case of one-way ANOVA, the two variables are the outcome variable and the grouping variable. We define the effect size V for one-way ANOVA with binary data as

$$V = \sqrt{-2 \sum_{j=1}^{k} w_j \left\{ \mu_j (\ln \mu_0 - \ln \mu_j) + (1 - \mu_j) \left[\ln(1 - \mu_0) - \ln(1 - \mu_j) \right] \right\} / (k - 1)},$$

where $w_j = n_j / n$ is the weight of the jth group, and $n = \sum_j^k n_j$ is the total size. The small, medium, and large effect size can be defined as 0.10, 0.30, and 0.50, borrowed from Cohen's effect size benchmarks (Cohen, 1988; Ellis, 2010).

7.3.1   Empirical effect size for one-way ANOVA with binary data

For a sample of data $Y = (y_j)$, $j = 1, 2, \cdots, k$, and $y_j = \{y_{ij}\}$, $i = 1, 2, \cdots, n_j$, where n_j is the sample size of the jth group, given the group sample size n_j and the group mean, as well as the proportion, $\bar{y}_j = \sum_i^{n_j} y_{ij} / n_j$, the sample effect size V for one-way ANOVA with binary data can be calculated as

$$\hat{V} = \sqrt{\bar{D} / n(k - 1)}$$

$$= \sqrt{-2 \sum_{j=1}^{k} n_j \left\{ \bar{y}_j (\ln \bar{y} - \ln \bar{y}_j) + (1 - \bar{y}_j) \left[\ln(1 - \bar{y}) - \ln(1 - \bar{y}_j) \right] \right\} / [n(k - 1)]}$$

with the grand mean obtained by $\bar{y} = \sum_j^k \bar{y}_j n_j / n$, the total sample size $n = \sum_j^k n_j$, and the weight $w_j = n_j / n$. A special case is for the balanced design where $n_1 = n_2 = \cdots = n_k = n/k$, which leads to $w_j = 1/k$ and the grant mean $\bar{y} = \sum_j^k \bar{y}_j / k$. Then

$$\hat{V} = \sqrt{-2 \sum_{j=1}^{k} \left\{ \bar{y}_j (\ln \bar{y} - \ln \bar{y}_j) + (1 - \bar{y}_j) \left[\ln(1 - \bar{y}) - \ln(1 - \bar{y}_j) \right] \right\} / [(k-1)k].}$$

The effect size can be determined based on the literature or empirical data. For example, one can obtain the information for each group in ANOVA as shown below and then use the above formulas to calculate the empirical effect size.

Group	Group size	Group Mean
1	n_1	$\bar{y}_1$
2	n_2	$\bar{y}_2$
$\vdots$	$\vdots$	$\vdots$
k	n_k	$\bar{y}_k$

To show how to calculate the effect sizes under different conditions, consider the following examples. For binary outcome, suppose the data are present as below.

	Group	Freshman	Sophomore	Junior	Senior
Balanced	Size	25	25	25	25
	Mean	.24	.28	.44	.56
Unbalanced	Size	24	30	26	20
	Mean	.24	.28	.44	.56

Table 7.3.1: Example data for effect size calculation: One-way ANOVA with binary data

For the balanced design, we first calculate the grant mean $\bar{y} = (.24 + .28 + .44 + .56)/4 = .38$, and then the effect size is

Example 7.3.1: Empirical effect size of the balanced design: One-way ANOVA with binary data

$$\begin{aligned} \hat{V} = \Big\{ &-2 \times \{.24 \times (\ln.38 - \ln.24) + (1 - .24) \times [\ln(1 - .38) - \ln(1 - .24)] \\ &+ .28 \times (\ln.38 - \ln.28) + (1 - .28) \times [\ln(1 - .38) - \ln(1 - .28)] \\ &+ .44 \times (\ln.38 - \ln.44) + (1 - .44) \times [\ln(1 - .38) - \ln(1 - .44)] \\ &+ .56 \times (\ln.38 - \ln.56) + (1 - .56) \times [\ln(1 - .38) - \ln(1 - .56)]\} / [(4 - 1) \times 4] \Big\}^{\frac{1}{2}} \\ =& 0.1530. \end{aligned}$$

For the unbalanced design, we first calculate the total sample size $n = 24 + 30 + 26 + 20 = 100$ and the grant mean $\bar{y} = (.24 \times 24 + .32 \times 30 + .04 \times 26 + .28 \times 20)/100 = .368$. Then the effect size is

Example 7.3.2: Empirical effect size of the unbalanced design: One-way ANOVA with binary data

$$\hat{V} = \left\{ -2/100 \times \{ 24 \times .24 \times (\ln .368 - \ln .24) + 24 \times (1 - .24) \times [\ln(1 - .368) - \ln(1 - .24)] \right.$$
$$+ 20 \times .28 \times (\ln .368 - \ln .28) + 20 \times (1 - .28) \times [\ln(1 - .368) - \ln(1 - .28)]$$
$$+ 26 \times .44 \times (\ln .368 - \ln .44 + 26 \times (1 - .44) \times [\ln(1 - .368) - \ln(1 - .44)]$$
$$\left. + 20 \times .56 \times (\ln .368 - \ln .56) + 20 \times (1 - .56) \times [\ln(1 - .368) - \ln(1 - .56)] \} / (4 - 1) \right\}^{\frac{1}{2}}$$
$$= 0.1465.$$

7.3.2   *Effect size calculator for one-way ANOVA with binary data*

The specification of the effect size can be assisted with an online calculator. For binary data, in the interface in Figure 7.5.1, clicking the link "Calculator" brings up the calculator. The calculator allows the calculation of effect sizes using two methods.

Method 1 allows a user to input group sample sizes and means (proportions) to calculate the effect size. By default, one can input data for three groups. However, one can specify any number of groups by inputting it in the *Number of groups* field and clicking the "Update" button to show more groups. For example, to obtain the effect sizes based on the unbalanced design in Table 7.3.1, the data can be input as in Figure 7.3.1. At the bottom, the effect size is presented after clicking the "Calculate" button.

Method 2 allows a user to upload a data file and calculates the effect size based on the data directly. Figure 7.3.2 shows the use of the raw data of Example 7.1.1 in Table 7.3.2 and the output of the effect size. Note that for the purpose of data privacy protection, only registered users are eligible for using this method. The data file has to be in text format where the first column of the data is the outcome variable named y and the second is the grouping variable named A. The first line of the data should be the variable names.

7.4   *Technical Details*

Let Y be a zero-one response/outcome variable, and A be a categorical variable of k levels. The null hypothesis H_0 assumes that samples in different groups are drawn from populations with equal proportions, while the alternative hypothesis H_1 supposes that at least two groups

Effect Size Calculator for Binary ANOVA
Method 1: Use group mean information

Number of groups: 4   Update

Group	Sample size	Mean
1	25	0.24
2	25	0.28
3	25	0.44
4	25	0.56

Calculate

Method 2: From empirical data analysis

Upload data file:
    Choose File | No file chosen   Calculate

Effect size output

The overall effect size   V = **0.1530**

Figure 7.3.1: Interface for effect size calculator of one-way ANOVA with binary data.

are from populations with different proportions. Let μ_j be the population proportion of the jth group, $j = 1, 2, \cdots, k$. The null and alternative hypotheses can be denoted as follows:

$H_0 :$ $\mu_1 = \mu_2 = \ldots = \mu_k = \mu_0,$
$H_1 :$ $\exists\ \mu_g \neq \mu_j,$ where $g \neq j$ and $g,\ j \in [1, 2, \cdots, k].$

Consider the corresponding models with H_0 and H_1. The null model M_0 is

$$E\{Y|A = j\} = \mu_0,$$

where $Y|(A = j) \sim Bernoulli(\mu_0)$, and the alternative model M_1 is

$$E\{Y|A = j\} = \mu_j,$$

where $Y|(A = j) \sim Bernoulli(\mu_j)$.

Given the sample data $Y = (y_j)$, $j = 1, 2, \cdots, k$, and $y_j = (y_{ij})$, $i = 1, 2, \cdots, n_j$, where n_j is the sample size of the jth group, the log-likelihood ratio of M_0 and M_1 is

Effect Size Calculator for Binary ANOVA
Method 2: From empirical data analysis

Upload data file:

Choose File | binary-anova-data.txt   Calculate

Effect size output

The overall effect size V = **0.153**

Output from Binary ANOVA

	Sum of Var	Df	Statistic	p-value
Between	7.02	3	7.02	0.929
Within	125.79	96		
Total	132.81			

$$
\begin{aligned}
\ln \frac{\mathcal{L}_{M_0}}{\mathcal{L}_{M_1}} &= \ln \frac{\mathcal{L}\left(\mu_0 \mid \boldsymbol{Y}\right)}{\mathcal{L}\left(\mu_1, \mu_2, \ldots \mu_k \mid \boldsymbol{Y}\right)} \\
&= \ln \frac{\prod_{j=1}^{k} \prod_{i=1}^{n_j} \mu_0^{y_{ij}}\left(1-\mu_0\right)^{1-y_{ij}}}{\prod_{j=1}^{k} \prod_{i=1}^{n_j} \mu_j^{y_{ij}}\left(1-\mu_j\right)^{1-y_{ij}}} \\
&= \sum_{j=1}^{k} \sum_{i=1}^{n_j}\left\{\left[y_{ij} \ln \mu_0+\left(1-y_{ij}\right) \ln \left(1-\mu_0\right)\right]-\left[y_{ij} \ln \mu_j+\left(1-y_{ij}\right) \ln \left(1-\mu_j\right)\right]\right\} \\
&= \sum_{j=1}^{k} \sum_{i=1}^{n_j}\left\{y_{ij}\left(\ln \mu_0-\ln \mu_j\right)+\left(1-y_{ij}\right)\left[\ln \left(1-\mu_0\right)-\ln \left(1-\mu_j\right)\right]\right\}.
\end{aligned}
$$

The log likelihood ratio test statistic

$$
\begin{aligned}
D &= -2 \ln \frac{\mathcal{L}\left(\mu_0 \mid \boldsymbol{Y}\right)}{\mathcal{L}\left(\mu_1, \mu_2, \ldots, \mu_k \mid \boldsymbol{Y}\right)} \\
&= -2 \sum_{j=1}^{k} \sum_{i=1}^{n_j}\left\{y_{ij}\left(\ln \mu_0-\ln \mu_j\right)+\left(1-y_{ij}\right)\left[\ln \left(1-\mu_0\right)-\ln \left(1-\mu_j\right)\right]\right\},
\end{aligned}
$$

follows a chi-square distribution $\chi^2(df)$, $df = k - 1$, according to the Wilks' theorem (Wilks, 1938).

Let the grand sample proportion $\bar{y} = \sum_j^k \sum_i^{n_j} y_{ij}/n$ be the estimate of μ_0, where $n = \sum_j^k n_j$ is the total sample size, and let the group proportion $\bar{y}_j = \sum_i^{n_j} y_{ij}/n_j$ be the estimate of μ_j, $j = 1, 2, \cdots, k$. For the given sample of data $\boldsymbol{Y}$ we can calculate the test statistic as

PRACTICAL STATISTICAL POWER ANALYSIS

y	A	y	A	y	A	y	A
0	1	0	2	0	3	1	4
0	1	0	2	1	3	0	4
0	1	1	2	1	3	1	4
0	1	0	2	0	3	1	4
1	1	0	2	1	3	0	4
0	1	0	2	1	3	1	4
0	1	1	2	0	3	1	4
0	1	0	2	0	3	0	4
1	1	0	2	0	3	0	4
0	1	0	2	0	3	1	4
0	1	0	2	0	3	0	4
0	1	0	2	0	3	0	4
1	1	1	2	0	3	1	4
0	1	0	2	0	3	1	4
1	1	0	2	0	3	1	4
0	1	0	2	1	3	1	4
0	1	1	2	1	3	0	4
0	1	1	2	1	3	1	4
0	1	0	2	1	3	1	4
1	1	0	2	0	3	0	4
0	1	0	2	1	3	1	4
0	1	1	2	1	3	0	4
1	1	0	2	0	3	0	4
0	1	0	2	0	3	1	4
0	1	1	2	1	3	0	4

Table 7.3.2: Raw data for effect size calculation for One-way ANOVA with binary data

Note. y is the success-failure records, A=1 is the freshman, A=2 is the sophomore, A=3 is the junior, and A=4 is the senior.

$$\tilde{D} = -2\ln \frac{\mathcal{L}\left(\bar{y}|Y\right)}{\mathcal{L}\left(\bar{y}_1, \bar{y}_2, \ldots, \bar{y}_k |Y\right)}$$

$$= -2\sum_{j=1}^{k}\sum_{i=1}^{n_j}\left\{ y_{ij}(\ln \bar{y} - \ln \bar{y}_j) + (1 - y_{ij})\left[\ln(1 - \bar{y}) - \ln(1 - \bar{y}_j)\right]\right\}$$

$$= -2\sum_{j=1}^{k} n_j \left\{ \bar{y}_j(\ln \bar{y} - \ln \bar{y}_j) + (1 - \bar{y}_j)\left[\ln(1 - \bar{y}) - \ln(1 - \bar{y}_j)\right]\right\}.$$

When the null hypothesis H_0 is true, the test statistic D follows a central chi-squared distribution $\chi^2(df)$, where $df = k - 1$ is the degree of freedom. If $\tilde{D}$ is larger than the critical value $C = \chi^2_{1-\alpha}(df)$ at the alpha level α, one would reject the null hypothesis H_0. When the alternative hypothesis H_1 is true, the test statistic D approximately follows a non-central chi-squared distribution $\chi^2(df, \lambda)$, where $df = k - 1$ is the degrees of freedom and $\lambda = n(k-1)V^2$ is the non-centrality parameter. Let $\Phi_{\chi^2(df,\lambda)}(x)$ be the cumulative distribution function of the non-central chi-square distribution. Then the statistical power of the test can be calculated as

$$
\begin{aligned}
power &= \Pr\{D \geq C \mid H_1\} \\
&= \Pr\{\chi^2(df,\lambda) \geq C\} \\
&= 1 - \Phi_{\chi^2(df,\lambda)}(C) \\
&= 1 - \Phi_{\chi^2[(k-1),n(k-1)V^2]}\left[\chi^2_{1-\alpha}(k-1)\right].
\end{aligned}
$$

7.5 How to Conduct Power Analysis for One-way ANOVA with Count Data

Figure 7.5.1 displayed the primary software interface for power analysis for one-way ANOVA with count data. Within the interface, a user can supply different parameter values and select different options for power analysis. Among the five parameters, *Number of groups*, *Sample size*, *Effect size*, *Significance level*, and *Power*, one and only one can be left blank. The usage of the interface for count data is similar to that for binary data. Some examples on how to use the interface are given below.

7.5.1 Examples

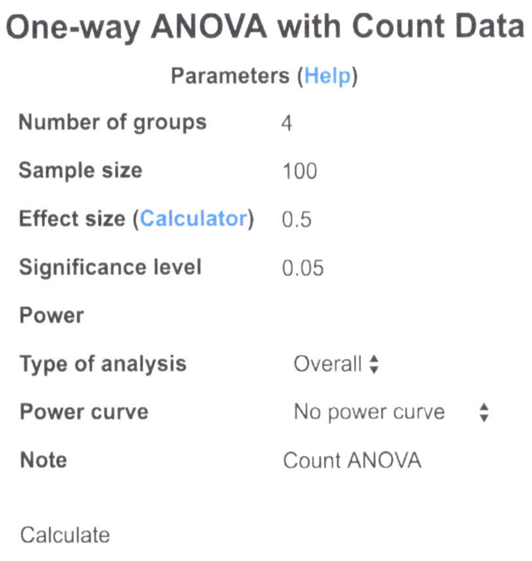

One-way ANOVA with Count Data

Parameters (Help)

Number of groups	4
Sample size	100
Effect size (Calculator)	0.5
Significance level	0.05
Power	
Type of analysis	Overall ⬍
Power curve	No power curve ⬍
Note	Count ANOVA

Calculate

Figure 7.5.1: Software interface of power analysis for one-way ANOVA with count data.

A local police officer in a downtown area hypothesizes that there are different numbers of crimes in the four seasons: spring, summer, fall, and winter. Based on his prior knowledge, he expects that the effect size is about 0.148. If he plans to investigate the criminal records from 25 years, what is the power for him to find the significant difference among the four seasons?

Example 7.5.1: Calculate power given sample size and effect size for one-way ANOVA with count data

The input and output for calculating power for this study are given in Figure 7.5.2. In the field of *Number of groups*, input 4, the total number of groups; in the field of *Sample size*, input 100, the total sample size of the four groups; and in the field of *Effect size*, input 0.148, the expected effect size. The default significance level 0.05 is used although one can change it to a different value. The field for *Power* is left blank because it will be calculated. By clicking the "Calculate" button, the statistical power is given in the output. For the current design, the power is 0.5597.

One-way ANOVA with Count Data

Figure 7.5.2: Input and output for calculating power for one-way ANOVA with count data in Example 7.5.1.

Parameters (Help)

Number of groups	4
Sample size	100
Effect size (Calculator)	0.148
Significance level	0.05
Power	
Type of analysis	Overall ⬍
Power curve	No power curve ⬍
Note	Count ANOVA

Calculate

Output

```
One-way Analogous ANOVA with Count Data

   k    n      V alpha  power
   4  100  0.148   0.05 0.5597


NOTE: n is the total sample size
URL: http://psychstat.org/anovacount
```

A power curve is a line plot of the statistical power along with given sample sizes. In Example 7.5.1, the power is 0.56 with the sample size 25 in each group. What is the power for a different sample size, say, 50 in each group? One can investigate the power of different sample sizes and plot a power curve.

The input and output for calculating power for the study in Example 7.5.1 with a sequence of sample sizes from 100 to 200 (25 to 50 each group) with an interval of 20 are given in Figure 7.5.3. Note that in the *Sample size* field, the input is 100:200:20. In the output, the power for each of the sample sizes is in the list. Especially, with the sample size 180 (the sample size is about 45 for each group), the power is about 0.8344. In the input, we also choose "Show power curve" in the *Power curve* drop-down menu. In the output, the power curve as shown in Figure 7.5.4 is displayed at the bottom of the output.

Example 7.5.2: Power curve

In practice, a power 0.8 is often desired. Given a power, the sample size can also be calculated as shown in Figure 7.5.5. In this situation, the *Sample size* field is left blank while in the *Power* field, the value 0.8 is provided. In the output, we can see a sample size 166 is needed to obtain a power 0.8, that is about 42 for each group.

Example 7.5.3: Calculate sample size given power and effect size

One can also calculate the minimum effect size to achieve a certain power given a sample size. As shown in Figure 7.5.6, we leave the *Effect size* field blank but provide input on *Sample size* (100) and *Power* (0.8). In the output, we can see the estimated minimum detectable effect size is 0.1906. This means that to get a power 0.8 with the sample size 100, the population effect size has to be at least 0.1906.

Example 7.5.4: Minimum detectable effect size

7.6   *Using R WebPower for One-way ANOVA with Count Data*

The online power analysis is carried out using the R package WebPower on a Web server. The package can be directly used within R for power analysis for one-way ANOVA with count data. Specifically, the function `wp.anova.count` is used. For example, the R input and output for Example 7.5.2 are given below. When the sample size is provided, the power will be calculated. The basic usage of the function is given below.

One-way ANOVA with Count Data

Parameters (Help)

Number of groups	4
Sample size	100:200:20
Effect size (Calculator)	0.148
Significance level	0.05
Power	
Type of analysis	Overall ↕
Power curve	Show power curve ↕
Note	Count ANOVA

Calculate

Figure 7.5.3: Input and output for power curve of one-way ANOVA with count data in Example 7.5.2.

Output

```
One-way Analogous ANOVA with Count Data

   k   n      V alpha  power
   4 100 0.148   0.05 0.5597
   4 120 0.148   0.05 0.6471
   4 140 0.148   0.05 0.7218
   4 160 0.148   0.05 0.7839
   4 180 0.148   0.05 0.8344
   4 200 0.148   0.05 0.8747


NOTE: n is the total sample size
URL: http://psychstat.org/anovacount
```

```
wp.anova.count(k = NULL, n = NULL, V = NULL, alpha = 0.05, power
    = NULL)
```

The examples in the previous section are included below.

```
> wp.anova.count(k = 4, n = 100, V = 0.148, alpha = 0.05, power =
    NULL)

Power for One-way ANOVA with Count Data

    k        n      V      alpha    power
    4       100    0.148   0.05    0.5597441

NOTE: n is the total sample size adding all groups (overall)
WebPower URL: http://psychstat.org/anovacount
```

V: effect size
k: number of groups
n: sample size
alpha: significance level
power: statistical power

Figure 7.5.4: Power curve of one-way ANOVA with count data in Example 7.5.2.

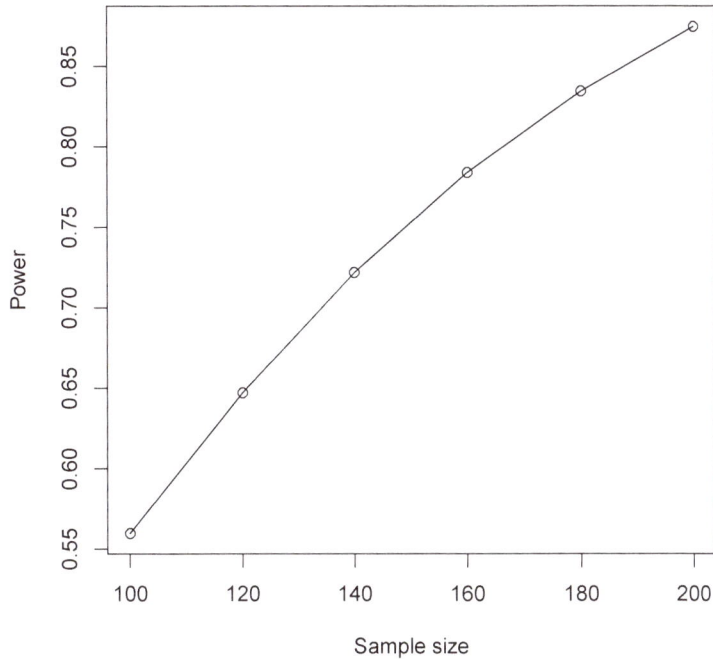

```
>## power curve
> powers <- wp.anova.count(k = 4, n = seq(100,200,10), V = 0.148,
    alpha = 0.05, power = NULL)
> powers

Power for One-way ANOVA with Count Data

        k       n       V       alpha   power
        4       100     0.148   0.05    0.5597441
        4       110     0.148   0.05    0.6049618
        4       120     0.148   0.05    0.6470911
        4       130     0.148   0.05    0.6860351
        4       140     0.148   0.05    0.7217782
        4       150     0.148   0.05    0.7543699
        4       160     0.148   0.05    0.7839101
        4       170     0.148   0.05    0.8105368
        4       180     0.148   0.05    0.8344142
        4       190     0.148   0.05    0.8557241
        4       200     0.148   0.05    0.8746580


NOTE: n is the total sample size adding all groups (overall)
WebPower URL: http://psychstat.org/anovacount
```

One-way ANOVA with Count Data

Figure 7.5.5: Input and output for sample size determination of one-way ANOVA with count data in Example 7.5.3.

Parameters (Help)

Number of groups	4
Sample size	
Effect size (Calculator)	0.148
Significance level	0.05
Power	0.8
Type of analysis	Overall ⬍
Power curve	No power curve ⬍
Note	Count ANOVA

Calculate

Output

```
One-way Analogous ANOVA with Count Data

    k     n      V alpha power
    4 165.9 0.148  0.05   0.8


NOTE: n is the total sample size
URL: http://psychstat.org/anovacount
```

```
> plot(powers)  ## generate power curve
```

Furthermore, we can also use R to estimate either the required sample size or the minimum detectable effect size.

```
> wp.anova.count(k = 4, n = NULL, V = 0.148, alpha = 0.05, power
    = 0.8)

Power for One-way ANOVA with Count Data

    k       n      V      alpha  power
    4       165.9143       0.148  0.05   0.8

NOTE: n is the total sample size adding all groups (overall)
WebPower URL: http://psychstat.org/anovacount


> wp.anova.count(k = 4, n = 100, V = NULL, alpha = 0.05, power =
    0.8)

Power for One-way ANOVA with Count Data
```

One-way ANOVA with Count Data

Parameters (Help)

Number of groups	4
Sample size	100
Effect size (Calculator)	
Significance level	0.05
Power	0.8
Type of analysis	Overall ⬍
Power curve	No power curve ⬍
Note	Count ANOVA

Calculate

Output

```
One-way Analogous ANOVA with Count Data


  k   n       V alpha power
  4 100 0.1906  0.05   0.8


NOTE: n is the total sample size
URL: http://psychstat.org/anovacount
```

Figure 7.5.6: Input and output for effect size calculation of one-way ANOVA with count data in Example 7.5.4.

```
    k      n      V        alpha    power
    4      100    0.1906373    0.05    0.8

NOTE: n is the total sample size adding all groups (overall)
WebPower URL: http://psychstat.org/anovacount
```

7.7   *Calculate Effect Size for One-way ANOVA with Count Data*

Similar to binary ANOVA analysis, we define the effect size V for one-way ANOVA with count data as

$$V = \sqrt{-2 * \sum_{j=1}^{k} w_j \left[\mu_j (\ln \mu_0 - \ln \mu_j) + (\mu_j - \mu_0) \right] / (k-1)},$$

where $w_j = n_j / n$ is the weight of the jth group, and $n = \sum_j^k n_j$ is the total size. The small, medium, and large effect size can be defined as

0.10, 0.30, and 0.50, borrowed from Cohen's effect size benchmarks (Cohen, 1988).

7.7.1   *Sample effect size for one-way ANOVA with count data*

For a sample of data $Y = (y_j)$, $j = 1,2,\cdots,k$, and $y_j = (y_{ij})$, $i = 1,2,\cdots,n_j$, where n_j is the sample size of the jth group. Let the group mean $\bar{y}_j = \sum_i^{n_j} y_{ij}/n_j$. The sample effect size V for one-way ANOVA with count data can be calculated as

$$\hat{V} = \sqrt{-2 \times \sum_{j=1}^{k} w_j \left[\bar{y}_j(\ln \bar{y} - \ln \bar{y}_j) + (\bar{y}_j - \bar{y}) \right] /(k-1)}$$

where $\bar{y} = \sum_j^k m_j n_j/n$ is the grand mean, $n = \sum_j^k n_j$ is the total sample size, and $w_j = n_j/n$ is the weight.

The effect size can be determined based on the literature or empirical data. For example, one can obtain the information for each group in ANOVA as shown below and then use the above formulas to calculate the sample effect size.

Group	Group size	Group Mean
1	n_1	$\bar{y}_1$
2	n_2	$\bar{y}_2$
$\vdots$	$\vdots$	$\vdots$
k	n_k	$\bar{y}_k$

We illustrate how to calculate the effect size using two examples with the information given in Table 7.7.1.

	Group	Freshman	Sophomore	Junior	Senior
Balanced	Group size	25	25	25	25
	Group mean	3.48	4.24	3.12	3.00
Unbalanced	Group size	30	24	26	20
	Group mean	3.48	4.24	3.12	3.00

Table 7.7.1: Example data for effect size calculation: One-way ANOVA with count data

For the balanced design, we first calculate the grand mean $\bar{y} = (3.48 + 4.24 + 3.12 + 3.00)/4 = 3.46$, and then the effect size

Example 7.7.1: Sample effect size of the balanced design for one-way ANOVA with count data

$$\hat{V} = \{ -2 \times [3.48 \times (\ln 3.46 - \ln 3.48) + (3.48 - 3.46)$$
$$+ 4.24 \times (\ln 3.46 - \ln 4.24) + (4.24 - 3.46)$$
$$+ 3.12 \times (\ln 3.46 - \ln 3.12) + (3.12 - 3.46)$$
$$+ 3.00 \times (\ln 3.46 - \ln 3.00) + (3.00 - 3.46)] / [(4-1) \times 4] \}^{\frac{1}{2}}$$
$$= 0.1480.$$

For the unbalanced design, we first calculate the total sample size $n = 30 + 24 + 26 + 20 = 100$ and the grand mean $\bar{y} = (3.48 \times 30 + 4.24 \times 24 + 3.12 \times 26 + 3.00 \times 20)/100 = 3.4728$. Then the effect size is

Example 7.7.2: Sample effect size of the unbalanced design for one-way ANOVA with count data

$$\hat{V} = \{ -2/100 \times [30 \times 3.48 \times (\ln 3.47 - \ln 3.48) + 30 \times (3.48 - 3.47)$$
$$+ 24 \times 4.24 \times (\ln 3.47 - \ln 4.24) + 24 \times (4.24 - 3.47)$$
$$+ 26 \times 3.12 \times (\ln 3.47 - \ln 3.12) + 26 \times (3.12 - 3.47)$$
$$+ 20 \times 3.00 \times (\ln 3.47 - \ln 3.00) + 20 \times (3.00 - 3.47)] / (4-1) \}^{\frac{1}{2}}$$
$$= 0.1427.$$

7.7.2   *Effect size calculator for one-way ANOVA with count data*

Two ways can be used to calculate the sample effect size. Method 1 allows a user to input group sample sizes and means to calculate the effect size. By default, one can input data for three groups. However, one can specify any number of groups by specifying it in the Number of groups field and clicking the "Update" button to show more groups. For example, to obtain the effect sizes based on the unbalanced design in Table 7.7.1, the data can be input as in Figure 7.7.2. At the bottom, the effect size for the overall effect size is presented after clicking on the "Calculate" button.

Method 2 allows a user to upload a set of data and calculates effect sizes based on the data directly. Figure 7.7.2 shows the use of the raw data of Example 7.5.1 in Table 7.7.2 and the output of the effect size. Note that only registered users can use this method to protect data privacy. The data file has to be in text format where the first column of the data is the outcome variable and the second is the grouping variable. The first line of the data should be the variable names.

Effect Size Calculator for Count ANOVA
Method 1: Use group mean information

Number of groups: 4   Update

Group	Sample size	Mean
1	25	3.48
2	25	4.24
3	25	3.12
4	25	3

Calculate

Method 2: From empirical data analysis

Upload data file:
    Choose File   No file chosen   Calculate

Effect size output

The overall effect size V = **0.1480**

7.8   *Technical Details*

Let Y be the type of data in which the observations can only take non-negative integer values, and A be a categorical variable of k levels. The null hypothesis H_0 assumes that samples in different groups are drawn from populations with equal means, while the alternative hypothesis H_1 assumes that at least two groups are from populations with different means. Let μ_j be the population mean of the jth group, $j = 1, 2, \cdots, k$, and μ_0 be the grant population mean. The null and alternative hypotheses can be denoted as follows:

$H_0: \quad \mu_1 = \mu_2 = \ldots = \mu_k = \mu_0,$

$H_1: \quad \exists \, \mu_g \neq \mu_j, \quad \text{where} \quad g \neq j \quad \text{and} \quad g, \, j \in [1, 2, \cdots, k].$

Consider the corresponding models with H_0 and H_1. The null model M_0 is

$$E\{Y|A = j\} = \mu_0,$$

where $Y|(A = j) \sim Poisson(\mu_0)$, and the alternative model M_1 is

$$E\{Y|A = j\} = \mu_j,$$

y	A	y	A	y	A	y	A
7	1	6	2	1	3	6	4
4	1	1	2	3	3	1	4
6	1	5	2	5	3	1	4
1	1	3	2	0	3	0	4
1	1	3	2	3	3	4	4
3	1	5	2	2	3	5	4
4	1	7	2	5	3	3	4
5	1	4	2	2	3	5	4
0	1	9	2	4	3	3	4
6	1	4	2	5	3	4	4
7	1	1	2	3	3	1	4
0	1	5	2	4	3	2	4
2	1	5	2	2	3	0	4
4	1	4	2	1	3	1	4
3	1	3	2	3	3	4	4
3	1	4	2	4	3	5	4
4	1	4	2	4	3	3	4
1	1	3	2	4	3	4	4
3	1	5	2	7	3	1	4
5	1	4	2	4	3	3	4
5	1	5	2	1	3	4	4
3	1	2	2	2	3	6	4
5	1	3	2	4	3	2	4
3	1	5	2	3	3	4	4
2	1	6	2	2	3	3	4

Table 7.7.2: Raw data for effect size calculation for one-way ANOVA with count data

where $Y|(A = j) \sim Poisson(\mu_j)$, $j = 1, 2, \cdots, k$.

Given the sample data $Y = (y_j)$, where $j = 1, 2, \cdots, k$, and $y_j = (y_{ij})$, $i = 1, 2, \cdots, n_j$, where n_j is the sample size of the jth group, the log likelihood ratio of M_0 and M_1 is

Effect Size Calculator for Count ANOVA
Method 2: From empirical data analysis

Figure 7.7.2: Input and output for effect
size calculator with empirical data.

Upload data file:

| Choose File | count-anova-data.txt | Calculate |

Effect size output

The overall effect size V = **0.148**

Output from Count ANOVA

```
        Sum of Var Df Statistic p-value
Between       6.567  3    6.567  0.9129
Within     -173.525 96
Total      -166.958
```

$$\ln \frac{\mathcal{L}_{M_0}}{\mathcal{L}_{M_1}} = \ln \frac{\mathcal{L}\left(\mu_0 | Y\right)}{\mathcal{L}\left(\mu_1, \mu_2, \ldots, \mu_k | Y\right)}$$

$$= \ln \frac{\prod_{j=1}^{k} \prod_{i=1}^{n_j} \left(\mu_0^{y_{ij}} e^{-\mu_0} / y_{ij}!\right)}{\prod_{j=1}^{k} \prod_{i=1}^{n_j} \left(\mu_j^{y_{ij}} e^{-\mu_j} / y_{ij}!\right)}$$

$$= \ln \frac{\prod_{j=1}^{k} \prod_{i=1}^{n_j} \left(\mu_0^{y_{ij}} e^{-\mu_0}\right)}{\prod_{j=1}^{k} \prod_{i=1}^{n_j} \left(\mu_j^{y_{ij}} e^{-\mu_j}\right)}$$

$$= \sum_{j=1}^{k} \sum_{i=1}^{n_j} y_{ij} \left(\ln \mu_0 - \mu_0\right) - \sum_{j=1}^{k} \sum_{i=1}^{n_j} y_{ij} \left(\ln \mu_j - \mu_j\right)$$

$$= \sum_{j=1}^{k} \sum_{i=1}^{n_j} \left[y_{ij}(\ln \mu_0 - \ln \mu_j) + (\mu_j - \mu_0)\right].$$

The log-likelihood ratio test statistic is

$$D = -2 * \ln \frac{\mathcal{L}\left(\mu_0 | Y\right)}{\mathcal{L}\left(\mu_1, \mu_2, \ldots, \mu_k | Y\right)}$$

$$= -2 * \sum_{j=1}^{k} \sum_{i=1}^{n_j} \left[y_{ij}(\ln \mu_0 - \ln \mu_j) + (\mu_j - \mu_0)\right],$$

which follows a chi-square distribution $\chi^2(df, \lambda)$, $df = k - 1$, and $\lambda \geq 0$, according to the Wilks' theorem (Wilks, 1938).

Let the grand mean $\bar{y} = \sum_{j}^{k} \sum_{i}^{n_j} y_{ij} / n$ be the estimate of μ_0, where $n = \sum_{j}^{k} n_j$ is the total sample size, and let the group mean $\bar{y}_j = \sum_{i}^{n_j} y_{ij} / n_j$ be the estimate of μ_j, $j = 1, 2, \cdots, k$. For a given sample of data Y we can calculate the test statistic as

ONE-WAY ANOVA WITH BINARY OR COUNT DATA

$$\tilde{D} = -2 * \ln \frac{\mathcal{L}\left(\bar{y}|Y\right)}{\mathcal{L}\left(\bar{y}_1, \bar{y}_2, \ldots, \bar{y}_k|Y\right)}$$

$$= -2 * \sum_{j=1}^{k} \sum_{i=1}^{n_j} \left[y_{ij}(\ln \bar{y} - \ln \bar{y}_j) + (\bar{y}_j - \bar{y}) \right]$$

$$= -2 * \sum_{j=1}^{k} n_j \left[\bar{y}_j(\ln \bar{y} - \ln \bar{y}_j) + (\bar{y}_j - \bar{y}) \right].$$

For one-way ANOVA with count data, we define the effect size as

$$V = \sqrt{-2 * \sum_{j=1}^{k} w_j \left[\mu_j(\ln \mu_0 - \ln \mu_j) + (\mu_j - \mu_0) \right] / (k-1)},$$

where $w_j = n_j/n$ is the weight of the jth group.

When the null hypothesis H_0 is true, the test statistic D follows a chi-squared distribution $\chi^2(df)$, where $df = k-1$ is the degrees of freedom. If $\tilde{D}$ is larger than the critical value $C = \chi^2_{1-\alpha}(df)$ at the alpha level α, one would reject the null hypothesis H_0. When the alternative hypothesis H_1 is true, the test statistic D approximately follows a non-central chi-squared distribution $\chi^2(df, \lambda)$, where $df = k-1$ is the degree of freedom and $\lambda = n(k-1)V^2$ is the non-centrality parameter. Let $\Phi_{\chi^2(df,\lambda)}(x)$ be the cumulative distribution function of the non-central chi-square distribution. Then the statistical power of the test can be calculated as

$$\begin{aligned} power &= \Pr\{D \geq C|H_1\} \\ &= \Pr\{\chi^2(df, \lambda) \geq C\} \\ &= 1 - \Phi_{\chi^2(df,\lambda)}(C) \\ &= 1 - \Phi_{\chi^2[(k-1), n(k-1)V^2]} \left[\chi^2_{1-\alpha}(k-1) \right]. \end{aligned}$$

7.9   Exercises

1. A researcher is planning to study the correct rate of classification with 4 groups: a memory intervention, a reasoning intervention, a processing speed intervention, and a control group. To decide the sample size, he has found the following proportion data from the literature. What would be the total sample size needed for a balanced design to get a power 0.8 at the alpha level 0.05?

	Memory	Reasoning	Speed	Control
Mean	.34	.40	.28	.33

2. Using the same information in Exercise 1, what would be the required sample sizes when the alpha level is set at 0.1 and 0.01, respectively?

3. Using the same information in Exercise 1, generate a power curve with the total sample size ranging from 100 to 2000 with an interval of 100. From the power curve, approximately how large a sample size is needed to get a power 0.9?

8   *Statistical Power Analysis for Repeated-Measures ANOVA*

Ge Jiang
Department of Educational Psychology
University of Illinois Urbana-Champaign

Repeated-measures ANOVA can be used to compare the means of a sequence of measurements. In a repeated-measures design, every subject is exposed to all treatment conditions, or more commonly, measured across different time points. If the subjects belong to one group, repeated-measures ANOVA can test the differences in the means of each condition, namely, the within-subject effects. If subjects are from more than one group, repeated-measures ANOVA can also test the differences between groups as well as the interaction effect between groups and measurements. The three tests use the F-statistics, where a large value implies that differences exist among the means of interest.

8.1   *How to Conduct Power Analysis for Repeated-Measures ANOVA*

The primary software interface for power analysis for repeated-measures ANOVA is shown in Figure 8.1.1. Within the interface, a user can supply different parameter values and select different options for power analysis. Among the four parameters, *Sample size*, *Effect size*, *Significance level*, and *Power*, one and only one can be left blank.

http://psychstat.org/rmanova

- The *Number of groups* tells how many groups, or how many levels of the between-subjects factor are used in the study design. At least one group is required.

- The *Number of measurements* tells how many repeated measurements, or how many levels of the within-subject factor are considered in the study design. At least two measurements are required, otherwise, regular ANOVA can be applied.

Repeated-Measures ANOVA

Figure 8.1.1: Software interface of power analysis for repeated-measures ANOVA

Parameters (Help)

Number of groups	4
Number of measurements	2
Sample size	100
Effect size (f) (Calculator)	0.5
Nonsphericity correction	1
Significance level	0.05
Power	
Type of effect	between effect
Power curve	No power curve
Note	Repeated-measures AN

Calculate

- The *Sample size* is the total number of participants from all groups. If the study design involves multiple groups of subjects, the equal sample size for each group is assumed for power calculation. The total sample size is then the product of the number of groups and the sample size for each group. For example, if 10 subjects are in each of the 3 groups, the total sample size is $3 \times 10 = 30$. Multiple total sample sizes can be provided in two ways to calculate power and produce a *power curve*. First, multiple sample sizes can be supplied and separated by white spaces, e.g., 100 150 200 will calculate power for the three sample sizes 100, 150 and 200. Second, a sequence of sample sizes can be generated using the method s:e:i with s denoting the starting sample size, e as the ending sample size, and i as the interval. Note that the values are separated by colon ":". For example, 100:150:10 will generate a sequence of sample sizes: 100 110 120 130 140 150. The default sample size, as shown in Figure 8.1.1, is 100.

- The *Effect size* specifies the magnitude of the population effect. Multiple effect sizes or a sequence of effect sizes can be supplied using the same way for sample size. By default, the value is 0.5. Determining the effect size is critical for power calculation. To help a user calculate effect sizes, a calculator has been developed and can be brought out by clicking the link "**Calculator**".

http://psychstat.org/rmanovaeffect

- The *Nonsphericity correction* specifies the degree of departure from sphericity. Repeated-measures ANOVA makes the assumption of sphericity about the population variance-covariance structure of the data. When this assumption is violated, a correction is required, called the non-sphericity correction. One can obtain some guidance on how to calculate this parameter by clicking the link "Nonsphericity". The default value is 1.0, implying that the sphericity assumption is met.

- The *Significance level*, or Type I error rate, for power calculation is needed and is usually set at the default level 0.05.

- The *Power* specifies the desired statistical power (usually set at 0.80) or can be left bank for calculation.

- The software can calculate power for between-subjects effects, within-subject effects and between-within interaction effects for repeated-measures ANOVA. The default analysis is for between-subjects effects. One can change that through *"Type of effect"*.

- In addition to the required input, one can also request the plot of a power curve if multiple sample sizes or effect sizes are provided. A note (less than 200 characters) can also be provided to add basic information on the analysis for future reference for registered users.

8.1.1   Examples

A researcher collects data from 30 toddlers by recording their age-normed general cognitive score at 30, 36, 42, and 48 months of age. The 30 children are raised in families with different socioeconomic status (low, middle, high). The researcher hypothesizes that the cognitive abilities of children with different SES are different (through the quality of powder the families can provide, the number of toys, etc.). Based on his prior knowledge, he expects that the effect size is about 0.36 and the nonsphericity correction is about 0.7. What is the power for this researcher to find a significant difference in cognitive abilities among children with different SES?

To compute power for this study, one should identify that age is the within-subject factor, SES is the between-subjects factor and the research question is a test of between-subjects effects. The input and output for calculating the power for this study are given in Figure 8.1.2. In the field of *Number of groups*, input 3, the total levels of SES; in the field of *Number of measurements,* input 4, the total number of time points; in the field of *Sample size*, input 30, the total sample size of all groups, and therefore, 10 in each SES group; in the field of *Effect size*, input 0.36, the expected effect size; and in the field of *Nonsphericity correction*, input 0.7, the assumed correction parameter. The default *Significance*

Example 8.1.1: Calculate power given sample size and effect size

level 0.05 is used although one can change it to a different value. The field for *Power* is left blank because it will be calculated. In the field of *Type of effect*, specify "between effect". By clicking the "**Calculate**" button, the statistical power is given in the output immediately. For the current design, the power is 0.2674.

To compute power for the test of within-subject effects or interaction effects, specify "within effect" or "interaction effect" in the field of *Type of effect*. Alter the value in *Effect size* if necessary and keep all other fields unchanged.

Repeated-Measures ANOVA

Parameters (Help)

Number of groups	3
Number of measurements	4
Sample size	30
Effect size (f) (Calculator)	0.36
Nonsphericity correction	0.7
Significance level	0.05
Power	
Type of effect	between effect ⬍
Power curve	No power curve ⬍
Note	Repeated-measures AN

Calculate

Output

```
Repeated-measures ANOVA analysis

   n     f ng nm nscor alpha  power
  30 0.36  3  4   0.7  0.05 0.2674


NOTE: Power analysis for between-effect test
URL: http://psychstat.org/rmanova
```

A power curve is a line plot of statistical power along with given sample sizes. In Example 8.1.1, the power is only 0.2674 with the sample size 10 in each group. What is the power for a different sample size, say, 40 in each group? One can investigate the power of different sample sizes and plot a power curve.

The input and output for getting a power curve for the study in

Example 8.1.1 are given in Figure 8.1.3. The sample size ranges from 30 to 150 (10 to 50 each group) with an interval of 30. In the *Sample size* field, the input is 30:150:30. We also choose "*Show power curve*" from the drop-down menu of *Power curve*. In the output, the power for each sample size from 30 to 150 with the interval 30 is listed. Particularly, with the total sample size 120 (40 for each group), the power is about 0.8389. The power curve is displayed at the bottom of the output (see Figure 8.1.4) and can be downloaded as a PDF file.

Repeated-Measures ANOVA

Figure 8.1.3: Input and output for power curve for repeated-measures ANOVA in Example 8.1.2

Parameters (Help)	
Number of groups	3
Number of measurements	4
Sample size	30:150:30
Effect size (f) (Calculator)	0.36
Nonsphericity correction	0.7
Significance level	0.05
Power	
Type of effect	between effect ⬍
Power curve	Show power curve ⬍
Note	Repeated-measures AN

Calculate

Output

```
Repeated-measures ANOVA analysis

     n     f ng nm nscor alpha  power
    30 0.36  3  4   0.7  0.05 0.2674
    60 0.36  3  4   0.7  0.05 0.5174
    90 0.36  3  4   0.7  0.05 0.7110
   120 0.36  3  4   0.7  0.05 0.8389
   150 0.36  3  4   0.7  0.05 0.9151

NOTE: Power analysis for between-effect test
URL: http://psychstat.org/rmanova
```

Example 8.1.3: Calculate sample size given power and effect size

Prior to data collection, a researcher might be interested to know how large a sample size is needed to obtain a certain level of power. In practice, a power of 0.8 is often desired. Given the power, the sample

Figure 8.1.4: Input and output for power
curve for repeated-measures ANOVA in
Example 8.1.2

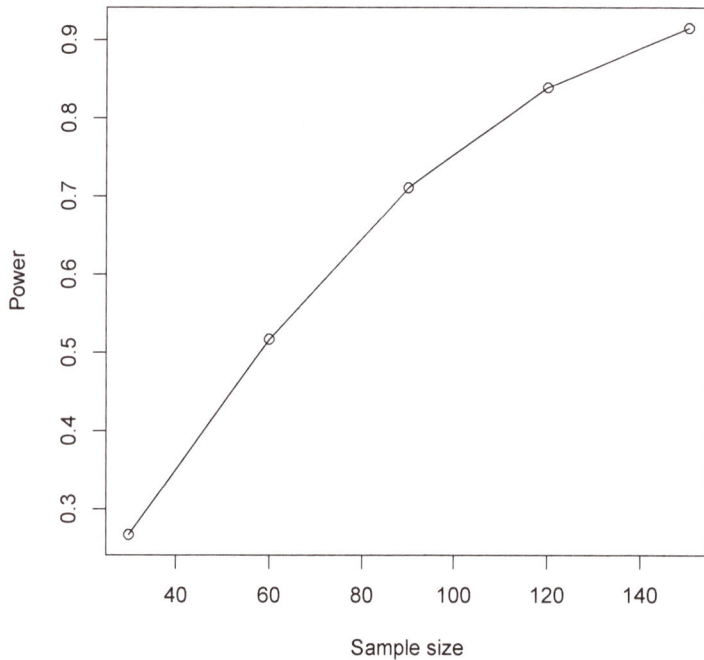

size can be calculated as shown in Figure 8.1.5. In this situation, the
Sample size field is left blank while the input for the *Power* field is 0.8.
In the output, we can see a total sample size 109.3 is needed to obtain
a power of 0.8. Since the sample size has to be an integer and also
divisible by 3, the minimum sample size required is 111, 37 for each
group.

A researcher may also be interested in knowing the magnitude of
the effect size that can be detected if he wants to maintain a certain
level of power. In Example 8.1.1, one can leave the field of *Effect size*
blank and input the desired power, often 0.8, in the field of *Power* as in
Figure 8.1.6. In the output, the effect size is 0.7168, meaning that given
the sample size and power, the minimum population effect size that
can be detected is 0.7168.

Example 8.1.4: Calculate effect size or
significance level

The software also provides the users the option to compute a significance level given sample size and power. It may be useful when the
control of type I error is less important than the control of power. In
Example 8.1.1, the input for *Power* is 0.8 and the field of *Significance
level* is left blank. In the output, the significance level is 0.4917.

Repeated-Measures ANOVA

Parameters (Help)

Number of groups	3
Number of measurements	4
Sample size	
Effect size (f) (Calculator)	0.36
Nonsphericity correction	0.7
Significance level	0.05
Power	0.8
Type of effect	between effect ⬍
Power curve	No power curve ⬍
Note	Repeated-measures AN

Calculate

Output

```
Repeated-measures ANOVA analysis

        n    f ng nm nscor alpha power
    109.3 0.36  3  4   0.7  0.05   0.8


NOTE: Power analysis for between-effect test
URL: http://psychstat.org/rmanova
```

Figure 8.1.5: Input and output for sample size planning for repeated-measures ANOVA in Example 8.1.3

8.2   Using R Package WebPower for Power Calculation

The power calculation for repeated-measures ANOVA is conducted using the R function wp.rmanova function. The detail of the function is:

```
wp.rmanova(n = NULL, ng = NULL, nm=NULL, f = NULL, nscor=1, alpha
    = 0.05, power = NULL, type=0)
```

The R input and output for the examples in the previous section are given below.

```
> ## power given sample size and effect size
> wp.rmanova(n=30, ng=3, nm=4, f=.36, nscor=.7)

Repeated-measures ANOVA analysis

    n    f ng nm nscor alpha     power
   30 0.36  3  4   0.7  0.05 0.2674167
```

n: sample size
ng: number of groups
nm: number of measurements
f: effect size
nscor: Nonsphericity correction coefficient
alpha: significance level
power: statistical power
type: type of analysis. 0, between-effect; 1, within-effect; 2, interaction effect

Repeated-Measures ANOVA

Figure 8.1.6: Input and output for deter-mining effect size for repeated-measures ANOVA in Example 8.1.4

Parameters (Help)

Number of groups	3
Number of measurements	4
Sample size	30
Effect size (f) (Calculator)	
Nonsphericity correction	0.7
Significance level	0.05
Power	0.8
Type of effect	between effect ↕
Power curve	No power curve ↕
Note	Repeated-measures AN

Calculate

Output

```
Repeated-measures ANOVA analysis

     n       f ng nm nscor alpha power
    30 0.7168  3  4   0.7  0.05   0.8


NOTE: Power analysis for between-effect test
URL: http://psychstat.org/rmanova
```

```
NOTE: Power analysis for between-effect test

WebPower URL: http://psychstat.org/rmanova

>
> ## power curve
> res <- wp.rmanova(n=seq(30, 150, 20), ng=3, nm=4, f=.36, nscor
    =.7)
> res

Repeated-measures ANOVA analysis

     n    f ng nm nscor alpha      power
    30 0.36  3  4   0.7  0.05  0.2674167
    50 0.36  3  4   0.7  0.05  0.4386000
    70 0.36  3  4   0.7  0.05  0.5894599
    90 0.36  3  4   0.7  0.05  0.7110142
```

```
    110 0.36  3  4   0.7  0.05 0.8029337
    130 0.36  3  4   0.7  0.05 0.8691834
    150 0.36  3  4   0.7  0.05 0.9151497

NOTE: Power analysis for between-effect test


WebPower URL: http://psychstat.org/rmanova


>
> plot(res)
>
> ## sample size given power and effect size
> wp.rmanova(n=NULL, ng=3, nm=4, f=.36, power=.8, nscor=.7)


Repeated-measures ANOVA analysis

            n     f ng nm nscor alpha power
     109.2546 0.36  3  4   0.7  0.05   0.8

NOTE: Power analysis for between-effect test


WebPower URL: http://psychstat.org/rmanova


>
> ## effect size given power and sample size
> wp.rmanova(n=30, ng=3, nm=4, f=NULL, power=.8, nscor=.7)


Repeated-measures ANOVA analysis

     n        f ng nm nscor alpha power
    30 0.716768  3  4   0.7  0.05   0.8

NOTE: Power analysis for between-effect test


WebPower URL: http://psychstat.org/rmanova
```

8.3   *Effect Size for Repeated-measures ANOVA*

We use the statistic f as the measure of effect size for repeated-measures ANOVA as in Cohen (1988, p. 275). The effect size can be calculated in similar ways for two-way ANOVA. The f is the ratio of the standard deviation σ_m of the effect to be tested and the within-cell standard deviation σ involved, multiplied by a coefficient such that

$$f = \frac{\sigma_m}{\sigma} * C. \tag{8.3.1}$$

The value of C is related to the effect to be calculated. Suppose J is the number of groups, K is the number of measurements and ρ is the correlation across repeated measurements. For between-subjects effects,

$$C = \sqrt{\frac{K}{1 + (K-1)\rho}},$$

and for within-subject effects and between-within interaction effects,

$$C = \sqrt{\frac{K}{1-\rho}}.$$

Given the three quantities σ_m, σ and C, the effect size can be determined as in Equation 8.3.1. Cohen defined the size of effect as: small 0.1, medium 0.25, and large 0.4 (Cohen, 1988).

The value of effect size can be obtained based on prior knowledge or calculated from empirical data in the literature. For example, the following cell means in Table 8.3.1 are obtained from an empirical repeated-measures study:

| | | \multicolumn{4}{c}{Measures} | |
		1	2	$\cdots$	K	Average
	1	μ_{11}	μ_{12}	$\cdots$	μ_{1K}	$\mu_{1\cdot}$
	2	μ_{21}	μ_{22}	$\cdots$	μ_{2K}	$\mu_{2\cdot}$
Group	$\vdots$	$\vdots$	$\vdots$	$\ddots$	$\vdots$	$\vdots$
	J	μ_{J1}	μ_{J2}	$\cdots$	μ_{JK}	$\mu_{J\cdot}$
Average		$\mu_{\cdot 1}$	$\mu_{\cdot 2}$	$\cdots$	$\mu_{\cdot K}$	$\mu_{\cdot\cdot}$

Table 8.3.1: Cell means from empirical studies

Note that μ_{jk}, $j = 1, 2, ..., J$, $k = 1, 2, ..., K$, are the means of the participants in group j at measurement k. The row means and column means are listed in the last column and the last row, respectively. With such information, the three quantities σ_m, σ, C, as well as the effect size f can be calculated. Assuming equal sample size for all groups, the between-group standard deviation σ_b can be calculated as

$$\sigma_b = \sqrt{\frac{\sum_{j=1}^{J} (\mu_{j\cdot} - \mu_{\cdot\cdot})^2}{J}},$$

where

$$\mu_{\cdot\cdot} = \sum_{j=1}^{J} \mu_{j\cdot}/J = \sum_{k=1}^{K} \mu_{\cdot k}/K$$

is the grand mean. Similarly, the standard deviation of the within-subject effects σ_w can be calculated as:

$$\sigma_w = \sqrt{\frac{\sum_{k=1}^{K} (\mu_{\cdot k} - \mu_{\cdot\cdot})^2}{K}}.$$

As for the interaction effects between the within-subject factor and the between-subjects factor, one first needs to compute the term $\mu_{jk} - \mu_{j\cdot} - \mu_{\cdot k} + \mu_{\cdot\cdot}$ for each cell, and then the standard deviation σ_i can be calculated as:

$$\sigma_i = \sqrt{\frac{\sum_{j=1}^{J}\sum_{k=1}^{K}(\mu_{jk} - \mu_{j\cdot} - \mu_{\cdot k} + \mu_{\cdot\cdot})^2}{JK}}.$$

The standard deviation in the denominator of Equation 8.3.1 is the common within-cell standard deviation σ, which can be calculated as:

$$\sigma = \sqrt{\frac{\sum_{j=1}^{J}\sum_{k=1}^{K}s_{jk}^2}{JK}},$$

where s_{jk}^2 denotes the variance among the participants within the corresponding cell.

The correlation ρ is typically calculated as the average of the $K(K-1)/2$ correlation coefficients among K repeated measurements. Given these quantities, the effect size is ready to be calculated. One can replace the numerator σ_m in 8.3.1 with σ_b, σ_w or σ_i depending on specific research questions.

8.3.1 Examples for effect size calculation

To show how to calculate the effect sizes for different effects, consider the following example. In a visual experiment, response time (measured in seconds) to a certain stimulus was recorded. The participants were from three groups (training groups 1, 2 and control group) with 30 participants in each group. Each participant was measured at 3 time points.

The table below summarizes the cell means in a similar way to Table 8.3.1. Suppose in the population, the common standard deviation is 2.66. Furthermore, we assume that the correlation across measurements is 0.1.

		Time points 1	2	3	Average
Groups	Training 1	13.2	11.4	10.4	11.67
	Training 2	16.8	12	5.8	11.53
	Control	11	9	8	9.33
Average		13.67	10.8	8.07	10.84

Table 8.3.2: Example data for effect size calculation

From the data, we have calculated the grand mean to be $\mu_{\cdot\cdot} = 1/3 \times (11.67 + 11.53 + 9.33) = 10.84$. The between group variance is

Example 8.3.1: Effect size for the between-subjects factor

$\sigma_b^2 = \sum_{j=1}^{J}(\mu_{j\cdot} - \mu_{\cdot\cdot})^2/J = [(11.67 - 10.84)^2 + (11.53 - 10.84)^2 + (9.33 - 10.84)^2]/3 = 1.15$, so the standard deviation is $\sigma_b = \sqrt{1.15} = 1.07$. The common within-cell standard deviation is $\sigma = 2.66$ and the correlation across measurements is $\rho = 0.1$. For the between-subjects effect size, the ratio of the two standard deviations needs to be multiplied by $C = \sqrt{K/[1+(K-1)\rho]} = \sqrt{3/[1+2*0.1]} = 1.58$. Therefore, the effect size is $f = (\sigma_b/\sigma)*C = (1.07/2.66)*1.58 = 0.6370$.

From the data, the within-subject variance is $\sigma_w^2 = \sum_{k=1}^{K}(\mu_{\cdot k} - \mu_{\cdot\cdot})^2/K = [(13.67 - 10.84)^2 + (10.8 - 10.84)^2 + (8.07 - 10.84)^2]/3 = 5.23$, so the standard deviation is $\sigma_w = \sqrt{5.23} = 2.29$. For the within-subject effect size, the coefficient to be multiplied is $C = \sqrt{K/(1-\rho)} = \sqrt{3/(1-0.1)} = 1.83$. Given $\sigma = 2.66$ and $\rho = 0.1$, the effect size is $f = (\sigma_w/\sigma)*C = (2.29/2.66)*1.83 = 1.5693$.

Example 8.3.2: Effect size for the within-subject factor

From the data, the term $\mu_{jk} - \mu_{j\cdot} - \mu_{\cdot k} + \mu_{\cdot\cdot}$ need to be first calculated for each cell. The variance for the interaction effects is $\sigma_i^2 = \sum_{j=1}^{J}\sum_{k=1}^{K}(\mu_{jk} - \mu_{j\cdot} - \mu_{\cdot k} + \mu_{\cdot\cdot})^2/JK = [(-1.29)^2 + (-0.22)^2 + 1.51^2 + 2.44^2 + 0.51^2 + (-2.96)^2 + (-1.16)^2 + (-0.29)^2 + 1.44^2]/9 = 2.50$ and the standard deviation is then $\sigma_i = \sqrt{2.50} = 1.58$. For the interaction effect size, the coefficient C to be multiplied is also 1.83. Therefore, the effect size is $f = (\sigma_i/\sigma)*C = (1.58/2.66)*1.83 = 1.0845$.

Example 8.3.3: Effect size for the interaction effects

8.3.2   Effect size calculator

The specification of the effect size can be assisted by an online calculator. Given the required parameters, the online calculator would produce the effect size accordingly. In the interface in Figure 8.3.1, clicking the link "**Calculate**" brings up the calculator. The calculator allows the calculation of effect sizes using four methods.

http://psychstat.org/rmanovaeffect

Method 1 (Figure 8.3.1) calculates the effect size based on *Cohen's f*. Given the required input, the effect size is calculated and shown at the bottom. The input for *Proportion of variance explained* is σ_m^2/σ^2 in Equation 8.3.1, the input for *number of measurements* is K and the input for *correlation across measurements* is ρ. By default, the values for the three parameters are 0.1, 1 and 0.5, respectively. However, the user can change the values in the boxes and click the "**Calculate**" button to update the result.

Example 8.3.4: Calculate effect size with method 1

Method 2 allows a user to input marginal means, error variance and the correlation across measurements to calculate the effect size of the two main effects. By default, one can input marginal means for three groups with each group measured three times. However,

Example 8.3.5: Calculate effect size with method 2

Effect Size Calculator for Repeated-Measures ANOVA

1. Effect size from explained variance

Proportion of variance explained   .1

Number of measurements   1

Correlation across measurements   .5

Calculate

Effect size output

The effect size f = **0.3333**

Figure 8.3.1: Effect size calculator for repeated-measures ANOVA

one can specify any number of groups and any number of levels of repeated measurements by inputting them and clicking the "**Update**" button. Note that when there is only one group, the calculator will only produce the effect size for the within-subject factor.

For example, to obtain the effect sizes for the data given in Table 8.3.2, the input for *Number of groups for factor A (between)* and *Number of groups for factor B (within, time)* are both 3. The inputs for *Factor A* are the marginal means, 11.67, 11.53 and 9.33, and the inputs for *Factor B* are 13.67, 10.8 and 8.07. The input for *Error variance* is $2.66^2 = 7.0756$. By default, the input for *Correlation across measurements* is 0.7, but one can change it to 0.1 in this example. By clicking the "**Calculate**" button, the effect sizes are presented at the bottom as in Figure 8.3.2. In the output, the effect size for factor A is $f = 0.6370$, and the effect size for factor B is $f = 1.5693$.

Method 3 allows a user to input cell means, error variance and the correlation across measurements to calculate three types of effect size. Similarly, the default number of groups and the number of repeated measurements are both three but one can specify any number of groups by changing the inputs and clicking the "**Update**" button.

For example, to obtain the effect sizes for the data given in Table 8.3.2, the input for *Number of groups for factor A (between)* and *Number of groups for factor B (within, time)* are both 3. The input for *A1* line are 13.2, 11.4 and 10.4, the input for *A2* line are 16.8, 12 and 5.8, and the input for *A3* line are 11, 9 and 8. The input for *Error variance* is 7.0756 and the input for *Correlation across measurements* is 0.1. By clicking the "**Calculate**" button, the effect sizes are presented at the bottom as in Figure 8.3.3. In the output, the effect size for factor A is $f = 0.6370$, the effect size for factor B is $f = 1.5693$, and the effect size for the interaction term is $f = 1.0845$.

Example 8.3.6: Calculate effect size with method 3

Example 8.3.7: Calculate effect size with method 4

Effect Size Calculator for Repeated-Measures ANOVA

Figure 8.3.2: Effect size calculation based on the input of marginal means

2. Use marginal mean information

Number of groups for factor A (between): 3
Number of groups for factor B (within, time): 3   Update

Factor A	11.67	11.53	9.33
Factor B	13.67	10.8	8.07
Error variance	7.0756		
Correlation across measurements	.1		

Calculate

Effect size output

The number of groups = 3
The number of measurements = 3
For Factor A
The effect size for Factor A: f = **0.6370**
For Factor B
The effect size for Factor B: f = **1.5693**

Method 4 allows a user to upload a data file and calculates effect sizes from the data directly. Figure 8.3.4 shows the use of the data in Table 8.3.2 and the output of the effect sizes. Note that only registered users can use this method to protect data privacy. In addition to the effect sizes, this method also conducts repeated-measure ANOVA analysis on the data uploaded.

The data file has to be in plain text format where the first column of the data is an index (id variable) column, the second is the outcome variable, the third is the measurement variable and the fourth is the grouping variable. If the first line of the data contains variable names, then one can choose *"With variable names"* from the drop-down menu of the *Data* tab. In the output, one can obtain the estimated correlation, estimated variance, and the three effect sizes. Results for repeated-measures ANOVA analysis and Mauchly's Test for Sphericity are also included. Note that the differences in effect size calculation are due to the use of estimated correlation and within-cell variance in Method 4. Given the same quantities, methods 2-4 should yield identical effect size estimates.

One such data set can be seen at http://psychstat.org/rmanovadata.

Effect Size Calculator for Repeated-Measures ANOVA

3. Use cell mean information

Number of groups for factor A (between): 3
Number of groups for factor B (within, time): 3   Update

	B1	B2	B3
A1	13.2	11.4	10.4
A2	16.8	12	5.8
A3	11	9	8
Error variance	7.0756		
Correlation across measurements	.1		

Effect size output

The number of groups = 3
The number of measurements = 3
For Factor A (between)
The effect size for Factor A: f = **0.6360**
For Factor B (within)
The effect size for Factor B: f = **1.5693**
For Interaction A*B
The effect size for A*B: f = **1.085**

8.4   *Technical Details*

The method is based on Muller et al. (1992) as used in Stata StataCorp (2013)

Repeated-measures designs are traditionally analyzed using ANOVA. When all assumptions are met, repeated-measures ANOVA is the most powerful method (O'Brien & Kaiser, 1985). The advantage is gained by the sphericity assumption that all repeated measures have equal variance and are equally correlated with each other. If the sphericity assumption is violated, approximated power can also be calculated by introducing the nonsphericity correction factor. WebPower primarily focuses on adjusted univariate approach with correction for nonsphericity.

Consider a two-way repeated-measures design where the between-subjects factor A has J levels and the within-subject factor B has K levels. A one-way repeated-measures design is a special case of two-way repeated-measures when $J = 1$. If there is more than one group, repeated-measures ANOVA assumes equal sample size n_0 for each group j, $j = 1, \ldots J$. The total sample size is $N = n_0 * J$.

There are three effects of interest: the two main effects of between-subject factor and within-subject factor, and the between-within interaction effects. The univariate tests of the three effects are easily computed once the corresponding multivariate tests have been computed. In multivariate approach, repeated-measures ANOVA is formed as a general linear multivariate model (GLMM):

$$\mathbf{Y} = \mathbf{XB} + \mathbf{E}, \tag{2}$$

where $\mathbf{Y}$ is a $N \times K$ matrix that contains the responses of N participants at K measurements, $\mathbf{X}$ is a $N \times J$ design matrix of rank J, $\mathbf{B}$ is a $J \times K$ matrix that contains all fixed effects and the $\mathbf{E}$ is a $N \times K$ error matrix with each row iid:

$$row_i(\mathbf{E}) \stackrel{d}{=} N_K(\mathbf{0}, \boldsymbol{\Sigma}).$$

In the previously discussed example of a visual experiment in section 8.3.1, the subjects are randomly assigned to $J = 3$ groups and each subject is measured at $K = 3$ time points, the GLMM can be expressed as:

$$\begin{pmatrix} y_{11} & y_{12} & y_{13} \\ y_{21} & y_{22} & y_{23} \\ \vdots & \vdots & \vdots \\ y_{N1} & y_{N2} & y_{N3} \end{pmatrix} = \begin{pmatrix} 1 & x_{11} & x_{12} \\ 1 & x_{21} & x_{22} \\ \vdots & \vdots & \vdots \\ 1 & x_{N1} & x_{N2} \end{pmatrix} \begin{pmatrix} \mu_1 & \mu_2 & \mu_3 \\ \alpha_{11} & \alpha_{12} & \alpha_{13} \\ \alpha_{21} & \alpha_{22} & \alpha_{23} \end{pmatrix} + \begin{pmatrix} \varepsilon_1' \\ \varepsilon_2' \\ \vdots \\ \varepsilon_N' \end{pmatrix}.$$

Here y_{ik} represents the response time (in seconds) of the ith subject measured at the kth time point. The first column in the design matrix $\mathbf{X}$ are all 1s in order to estimate the mean effects. The remaining columns of the design matrix are filled with dummy coding of the between-subjects factor:

$$x_{i1} = \begin{cases} 1 & \text{if subject } i \text{ is in training group 1} \\ 0 & \text{if subject } i \text{ is in training group 2} \\ -1 & \text{if subject } i \text{ is in control group} \end{cases}$$

and

$$x_{i2} = \begin{cases} 0 & \text{if subject } i \text{ is in training group 1} \\ 1 & \text{if subject } i \text{ is in training group 2} \\ -1 & \text{if subject } i \text{ is in control group} \end{cases}$$

The elements in $\mathbf{B}$ matrix are all fixed effects and will be used to form the hypotheses. The μ_k in the top row are the mean response time of all subjects at time period k, $k = 1, \ldots 3$. The α_{jk} represents the effect of jth group, $j = 1, 2$, at time period k, and $\alpha_{3k} = -\alpha_{1k} - \alpha_{2k}$. The ε_i' is a vector that contains the errors of the ith subject at the three time periods.

8.4.1   *Hypothesis testing*

The usual null hypothesis in the multivariate model involves the secondary parameter $\mathbf{\Theta} = \mathbf{CBU}$:

$$H_0: \ \mathbf{\Theta} = 0$$

where $\mathbf{\Theta}$ is a $a \times b$ matrix whose dimensions depends on the two contrast matrices $\mathbf{C}$ and $\mathbf{U}$. The matrix $\mathbf{C}$ is a $a \times J$ matrix of full row rank and $\mathbf{U}$ is a $K \times b$ matrix of full column rank. Each row of $\mathbf{C}$ defines a row of $\mathbf{\Theta}$ and forms a contrast among the between-subjects effects. Each column of $\mathbf{U}$ defines a column of $\mathbf{\Theta}$ and forms a contrast among the within-subject effects. Together, $\mathbf{C}$ and $\mathbf{U}$ can be used to test the between-within interaction effects. The contrasts in $\mathbf{C}$ and $\mathbf{U}$ can be arbitrarily changed to accommodate different hypotheses.

Tests of the hypothesis are typically based on the following estimates:

$$\hat{\mathbf{B}} = (\mathbf{X}'\mathbf{X})^{-1}\mathbf{X}'\mathbf{Y}$$

$$\hat{\mathbf{\Theta}} = \mathbf{C}\hat{\mathbf{B}}\mathbf{U}$$

$$\hat{\mathbf{H}} = \hat{\mathbf{\Theta}}' \left\{ \mathbf{C}(\mathbf{X}'\mathbf{X})^{-1}\mathbf{C}' \right\}^{-1} \hat{\mathbf{\Theta}}$$

and

$$\hat{\mathbf{E}} = \mathbf{U}'\hat{\mathbf{\Sigma}}\mathbf{U}(N-r)$$

with $r = rank(X)$. Note that in GLMM, the design matrix $\mathbf{X}$ might not be full rank. When $rank(X) = r < min(N, J)$, the generalized inverse $(X'X)^-$ is used. In any of the two following conditions: (1) Compound symmetry holds and $\mathbf{U}$ is orthonormal matrix, or (2) Sphericity holds, a usual test statistic for testing the null hypothesis is given by:

> Sphericity and Compound symmetry will be explained and compared later.

$$F_{obs} = \frac{\text{tr}(\hat{\mathbf{H}})/ab}{\text{tr}(\hat{\mathbf{E}})/b(N-r)},$$

which follows an exact F distribution with degrees of freedom of ab and $b(N-r)$ under the null. If all assumptions are met except sphericity, the distribution of the F_{obs} can be approximated by an adjusted F distribution:

$$F_{obs} \xrightarrow{d} F_{[ab\varepsilon, \ b(N-r)\varepsilon]}.$$

The ε can take values between $1/b$ and 1. When sphericity holds, $\varepsilon = 1$, if not, $\varepsilon < 1$, thus often referred to as nonspericity correction. The correction protects the test from being too liberal by reducing the numerator and denominator degrees of freedom and increasing the critical value. Two viable approaches for computing empirical ε are suggested for researchers. One is proposed by Greenhouse & Geisser (1959):

$$\hat{\varepsilon} = \frac{\text{tr}^2(\hat{\boldsymbol{\Sigma}})}{b\,\text{tr}(\hat{\boldsymbol{\Sigma}}^2)} = \frac{\left[\Sigma_{k=1}^{b}\lambda_k\right]^2}{b\,\Sigma_{k=1}^{b}\lambda_k^2}, \tag{3}$$

where λ_k, $k = 1, 2, ..., b$ are the ordered eigenvalues of $\hat{\boldsymbol{\Sigma}}$. The other is suggested by Huynh & Feldt (1976):

$$\tilde{\varepsilon} = \frac{Nb\hat{\varepsilon} - 2}{b(N - r - b\hat{\varepsilon})}, \tag{4}$$

which is based on Greenhouse-Geisser's $\hat{\varepsilon}$.

Under the alternative hypothesis, the distribution of F_{obs} is approximated by a noncentral F distribution with df of $ab\varepsilon$ and $b(N - r)\varepsilon$ with the noncentrality parameter λ as:

$$\lambda = ab\varepsilon F_{obs} = N\varepsilon f^2, \tag{5}$$

with f denoting the effect size defined in Section 8.3.

8.4.2 Power computation

Given the relevant quantities, power can be readily calculated for a specific test by referring F_{obs} to the approximate non-central F distribution under the alternative:

$$F_{obs} \xrightarrow{d} F_{[ab\varepsilon,\, b(N-r)\varepsilon,\, \lambda]}.$$

Specifically, the power (π) can be calculated as:

$$\pi = 1 - F_{ab\varepsilon, b(N-r)\varepsilon, \lambda}\left(F^{-1}_{ab\varepsilon, b(N-r)\varepsilon, 1-\alpha}\right) \tag{6}$$

where $F_{df_1,\, df_2,\, \lambda}$ is the cumulative distribution function of a non-central F and $F^{-1}_{df_1,\, df_2,\, 1-\alpha}$ gives the critical value of an F distribution given the tail probability $1 - \alpha$.

8.4.3 Sphericity

One of the major distinctions between repeated-measures designs and two-way factorial designs is that the repeated measures are often correlated. The measurements between subjects are still independent, but the measurements within a subject are often correlated. Therefore, to use the traditional univariate ANOVA approach to analyze repeated-measures data, the data must meet a set of rather restrictive assumptions. Besides the usual assumptions of random sampling from the population, independence of subjects, and normality, repeated-measures ANOVA assumes that the population covariance matrix has a certain form.

This form, which is called *sphericity* (or interchangeably, *circularity*), is the condition where the variances of the differences between all combinations of related treatments are equal. This means that if we take any two treatment levels, and subtract scores for one level from scores for another level, the resulting score must have the same population variance for every pair of levels.

Violation of sphericity is when the variances of the differences between all pairs of related measures are not equal. Failure to meet the assumption can cause the F-test too liberal and increase the Type I error rate. A special case of this assumption is compound symmetry, a less stringent one. How well this assumption is met can be formally tested and if the violations of sphericity do occur, corrections have been proposed.

8.4.4   *Compound symmetry*

One form of sphericity that is often discussed is called compound symmetry. A covariance matrix is defined to possess compound symmetry if and only if all variances of the repeated measurements are equal to each other and all the covariances are equal to each other. An equivalent property is that every measure has the same variance and all pairwise correlations are equal.

Note that when compound symmetry holds, the variances and covariances are constant regardless of the levels of the treatments. Thus, the variances of the differences between all pairs of related treatments are equal. Compound symmetry implies that the sphericity assumption is satisfied. However, compound symmetry is a sufficient condition but not a necessary condition because it is a special case of sphericity. Though it is not necessary, the absence of compound symmetry does indicate that sphericity is unlikely.

Matrices that satisfy compound symmetry are a subset of those that satisfy sphericity. In practice, there is only one related situation in which the distinction between sphericity and compound symmetry is of potential importance. When the within-subject factor has only two levels, there is just one difference between levels, so sphericity is always satisfied. However, the population covariance matrix does not necessarily possess compound symmetry because the variance at level 1 may not equal the variance at level 2. Besides this condition, it would be highly unusual (although theoretically possible) to find a matrix that possesses sphericity but not compound symmetry. Thus, for practical purposes, compound symmetry is always required for repeated-measures ANOVA any time the within-subject factor has more than two levels.

Sphericity test and nonsphericity correction

In most psychological studies, one would expect measurements taken prior to a treatment to correlate more highly with one another than with those taken after the treatment. Even in cases with no active treatment, one would expect successive or adjacent measurements to covary more highly than non-adjacent measurements. Clearly, the assumption of equal variances and equal pairwise correlations is often unrealistic and sphericity is unlikely to be valid under such conditions.

Mauchly's test of sphericity is a formal way of testing the sphericity assumption (Mauchly, 1940). The test is conducted in the same way as hypothesis testing. If the associated p-value is greater than or equal to the significance level, we could conclude that the assumption has not been violated. It is commonly used although O'Brien & Kaiser (1985) noted that such test can be quite sensitive to violations of normality as well as small sample size.

When the sphericity assumption is not met, a variety of procedures, generally involving reduction of the degrees of freedom through multiplication by some value ε, have been suggested to protect against liberal F test. The procedures differ in the way they compute the nonsphericity correction factor ε, which can be seen as a measure of the degree of sphericity in the population. An ε of 1 means sphericity is met. If the assumption is not met, $\varepsilon < 1$, and smaller value means a further departure from sphericity. The lowest value of ε is $1/(K-1)$ where K is the total number of measurements.

8.5   *Exercises*

1. A developmental psychologist is interested in the role of the sound of a mother's heartbeat in the growth of newborn babies. Fourteen babies were randomly selected and placed in the nursery. Specifically, the first seven babies were exposed to a rhythmic heartbeat sound piped in over the PA system. The other seven babies were placed in an identical nursery, but without the heartbeat sound. Infants were weighed at the same time of day for four consecutive days, yielding the following data (weight is measured in ounces):

    What is the sample effect size of the between-subjects factor (heartbeat sound)? Assuming the population effect size is identical to the sample effect size, what is the power for the psychologist to find a significant difference in weights between infants exposed to mother's heartbeat sound and the infants who are not using the sample effect size as the population effect size?

This example is based on the exercise of Maxwell & Delaney (2003, p. 491)

	Day 1	Day 2	Day 3	Day 4
1	96	98	103	104
2	116	116	118	119
3	102	102	101	101
4	112	115	116	118
5	108	110	112	115
6	92	95	96	98
7	120	121	121	123
8	112	111	111	109
9	95	96	98	99
10	114	112	110	109
11	99	100	99	98
12	124	125	127	126
13	100	98	95	94
14	106	107	106	107

2. Using the same information in Exercise 1, what is the sample effect size of the within-subject factor? Assuming the population effect size is the same as the sample effect size, what is the power for the psychologist to find a significant difference in weights between different days?

3. If the researcher can collect more data for both groups, what would be the power for a test of within-subject effects with a total of 100 participants?

4. Using the same information in Exercise 1, generate a power curve with the total sample size ranging from 20 to 200 with an interval of 10 for the test of within-subject effects. From the power curve, approximately how large is a sample size needed to get a power 0.9?

5. Using the same information in Exercise 1, if the researcher is interested in the interaction effects between the heartbeat sound and the days, what would be the sample size required to detect a significant difference with a power of 0.8 (Assume the population effect size the same as the sample effect size)?

6. Using the same information in Exercise 1, what would be the required sample sizes when the alpha level is set at 0.1 and 0.01, respectively?

Figure 8.3.4: Effect size calculation based on empirical data for repeated-measures ANOVA

Effect Size Calculator for Repeated-Measures ANOVA

4. From empirical data analysis

Upload data file:

Choose File | rm-anova-effect-data.txt   Calculate

Data   With variable names ⬍

Effect size output

The number of groups = **3**
The number of measurements = **3**
The estimated correlation across measurements = **-0.06836784**
The estimated error variance = **7.1**
For Factor A (between)
The effect size for Factor A: f = **0.7485**
For Factor B (within)
The effect size for Factor B: f = **1.4379**
For Interaction A*B
The effect size for A*B: f = **0.9937**

Output from Repeated-measures ANOVA

```
Error: id
          Df Sum Sq Mean Sq F value Pr(>F)
g          2  51.51  25.756   5.117 0.0247 *
Residuals 12  60.40   5.033
---
Signif. codes:  0 '***' 0.001 '**' 0.01 '*' 0.05 '.' 0.1 ' ' 1

Error: Within
          Df Sum Sq Mean Sq F value   Pr(>F)
time       2  235.2  117.62  16.567 3.02e-05 ***
time:g     4  112.4   28.09   3.956   0.0132 *
Residuals 24  170.4    7.10
---
Signif. codes:  0 '***' 0.001 '**' 0.01 '*' 0.05 '.' 0.1 ' ' 1
```

Mauchly's Test for Sphericity

The Mauchly's test statistic is 0.9331 with p-value 0.6834
The Greenhouse-Geisser estimate of the spheericity epsilon = 0.9373

9 | *Statistical Power Analysis for Linear Regression*

Agung Santoso
Department of Psychology
University of Notre Dame

Regression is a statistical technique for examining the relationship between one or more independent variables and one dependent variable. The independent variables are often called predictors or covariates, while the dependent variable is also called an outcome variable or criterion. Although regression analysis is commonly used to test the linear relationship between continuous predictors and outcome, it may also be used to test the interaction between predictors that are either continuous or categorical by utilizing dummy or contrast coding.

Regression provides an F-statistic that can be formulated using the ratio between the variance of the outcome variable explained by the predictors and the unexplained variance (Maxwell & Delaney, 2003):

$$F = \frac{SS_{regression}/df_{regression}}{SS_{residual}/df_{residual}} \tag{9.0.1}$$

where $SS_{regression}$ is the sum of squares of the outcome variable that is explained by the predictors and $SS_{residual}$ is the sum of squares of the residuals. F-statistic can also be expressed in terms of comparison between Full and Reduced models (Maxwell & Delaney, 2003):

$$F = \frac{(SSE_{Reduced} - SSE_{Full})/(dfE_{Reduced} - dfE_{Full})}{(SSE_{Full}/dfE_{Full})} \tag{9.0.2}$$

where $SSE_{Reduced}$ is the unexplained variation of the reduced model and SSE_{Full} is the unexplained variation of the full model. The reduced model is the one that has more parameters constrained than the full model. The reduced model is nested within the full model.

The second expression is very useful in testing the effect of a set of independent variables on an outcome variable controlling for the other variables in the model. The effect of a set of independent variables given the controlling variables can be expressed as the difference between the full model and the reduced model. Here, the reduced model consists of only the outcome variable and the controlling variables, while the full model also includes the independent variables. In other words, the parameters of the independent variables of interest in the reduced model are constrained to be zero, while in the full model, they are freely estimated.

9.1 How to Conduct Power Analysis for Linear Regression

The primary software interface for power analysis for linear regression analysis is shown in Figure 9.1.1. There are several parameters in the interface, namely *Sample size*, *Number of predictors of the full model*, *Number of predictors of the reduced model*, *Effect size*, *Significance level*, and *Power*. Only one field of *Sample size*, *Effect size*, or *Power* can be left blank for each analysis depending on the interest of analysis.

Linear Regression

Figure 9.1.1: Software interface for power analysis for linear regression

Parameters (Help)

Sample size	100
Number of predictors	
Full model	1
Reduced model	0
Effect size Show	0.15
Significance level	0.05
Power	
Power curve	No power curve ⬍
Note	Linear regression

Calculate

- The *Sample size* is the number of participants in the study. Multiple sample sizes can be provided to calculate power corresponding to each of them. Two ways of inputting multiple sample sizes can be used. First, multiple sample sizes can be typed, separated by white space, e.g. `100 150 200`, which will calculate power for three sample sizes 100, 150 and 200. Second, a sequence of sample sizes can be generated using the command `s:e:i` with `s` denoting the starting sample size, `e` as the ending sample size, and `i` as the interval. For example, `100:150:10` will generate a sequence of sample sizes of

100, 110, 120, 130, 140, and 150. The command will provide the same result as inputting `100 110 120 130 140 150` but in a more convenient way. The default sample size, as shown in Figure 9.1.1, is 100.

- The *Number of predictors* for both the full model and the reduced model is needed. By default, the number of predictors for the reduced model is 0.

- The *Effect size* specifies the size of the relationship between the predictors and the criterion in term of f^2. Multiple effect sizes or a sequence of effect sizes can also be supplied using the same method as for the *Sample size*. The default value for the *Effect size* is .15.

- The *Significance level* is the type I error rate that will be used to test the null hypothesis. Its default value is set to .05.

- *Power* specifies the desired statistical power.

- In addition to the required input, one can also request the plot of a power curve if multiple sample sizes or effect sizes are provided. A note that is less than 200 characters can also be provided as the basic information on the analysis for the future reference for registered users.

9.1.1 *Examples*

A researcher wants to examine the effect of intelligence, hours of study at home and parental support on students' achievement. She hypothesizes that at least one of the three predictors have a significant relationship with students' achievement. The type I error rate that she allows to test the hypothesis is .05. The expected effect size is .10 based on her literature review. She would like to know the power of the analysis if she collects data from 100 students.

Example 9.1.1: Calculate power given sample size and effect size

The input and output for calculating power for this study are given in Figure 9.1.2. In the field of *Sample size*, input `100`, the number of students in the study; in the field of *Number of predictors*, input `3` for the full model and `0` for the reduced model; and in the field of *Effect size*, input `0.10`, the expected effect size. The significance level is 0.05. The field for *Power* is left blank because it will be calculated. By clicking the "**Calculate**" button, the statistical power is given in the output. For the current analysis, the power is `0.742`.

If a researcher needs to look at the power of more than one sample size to make a decision, he or she may consult a power curve. For example, if the researcher in Example 9.1.1 is also interested in examining the

Example 9.1.2: Power curve

Linear Regression

Parameters (Help)

Sample size	100
Number of predictors	
Full model	3
Reduced model	0
Effect size Show	0.1
Significance level	0.05
Power	
Power curve	No power curve ⬍
Note	Linear regression

Calculate

Output

```
Power for multiple regression

    n p1 p2  f2 alpha power
  100  3  0 0.1  0.05 0.742

URL: http://psychstat.org/regression
```

power of the analysis across sample sizes ranging from 50 to 300 with an interval of 50, she can investigate the power of the desired sample sizes and plot a power curve. The input and output for calculating power and producing the power curve for this purpose can be seen in Figure 9.1.3.

Note that in the *Sample size* field, the input is `50:300:50`. In the output, the power for each sample size from 50 to 300 with the interval 50 is listed. The researcher can see that, for example, the power of the analysis by using only 50 respondents is `.4078`, while using 300 respondents the power will increase to `.9981`.

We may ask the software to show a power curve by choosing "*Show power curve*" in the drop-down menu of *Power curve*. The power curve is displayed in Figure 9.1.4. The power curve can be used for interpolation. For example, to obtain a power of .8, the researcher will need about 120 respondents. She can also see that adding more respondents will not increase the power of the analysis substantially after the sample size reaches 150.

Linear Regression

Figure 9.1.3: Input and output for calculating power and producing power curve for regression in Example 9.1.2

Parameters (Help)

Sample size   50:300:50

Number of predictors

  Full model   3

    Reduced model   0

Effect size Show   0.1

Significance level   0.05

Power

Power curve   Show power curve ⬍

Note   Linear regression

Calculate

Output

```
Power for multiple regression

    n p1 p2  f2 alpha  power
   50  3  0 0.1  0.05 0.4078
  100  3  0 0.1  0.05 0.7420
  150  3  0 0.1  0.05 0.9092
  200  3  0 0.1  0.05 0.9725
  250  3  0 0.1  0.05 0.9925
  300  3  0 0.1  0.05 0.9981


URL: http://psychstat.org/regression
```

One can also obtain the sample size given the desired power. In this case, we left the *Sample size* field blank, and input .8 in the *Power* field as shown in Figure 9.1.5. We can see that, in the output, the sample size needed to obtain the power of .8 is 113.

9.2   *Using R Package WebPower for Power Analysis for Linear Regression*

The power calculation for the linear regression is conducted using the R function wp.regression. The detail of the function is:

```
wp.regression(n = NULL, p1 = NULL, p2 = 0, f2 = NULL, alpha =
    0.05, power = NULL)
```

The R input and output for some examples discussed earlier are

n: sample size
p1: number of predictors in the full model
p2: number of predictors in the reduced model, 0 by default
f2: effect size
alpha: significance level
power: statistical power

Figure 9.1.4: Power curve for a linear regression analysis in Example 9.1.2

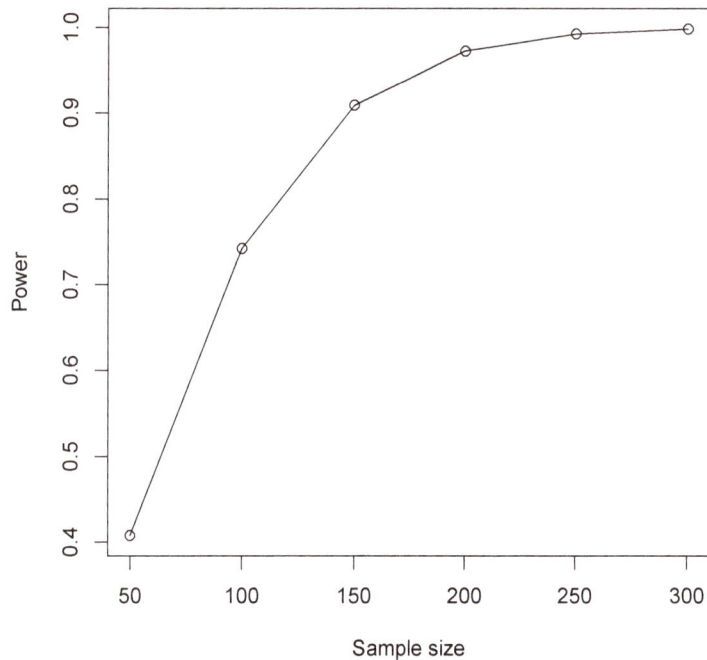

given below.

```
> ## calculate power given sample size and effect size
> wp.regression(n=100, p1=3, f2=.1)

Multiple regression power calculation

    n p1 p2  f2 alpha     power
  100  3  0 0.1  0.05 0.7420463

WebPower URL: http://psychstat.org/regression

>
> ## power curve
> res <- wp.regression(n=seq(50, 300, 50), p1=3, f2=.1)
> res

Multiple regression power calculation

    n p1 p2  f2 alpha     power
   50  3  0 0.1  0.05 0.4077879
  100  3  0 0.1  0.05 0.7420463
  150  3  0 0.1  0.05 0.9092082
  200  3  0 0.1  0.05 0.9724593
  250  3  0 0.1  0.05 0.9925216
```

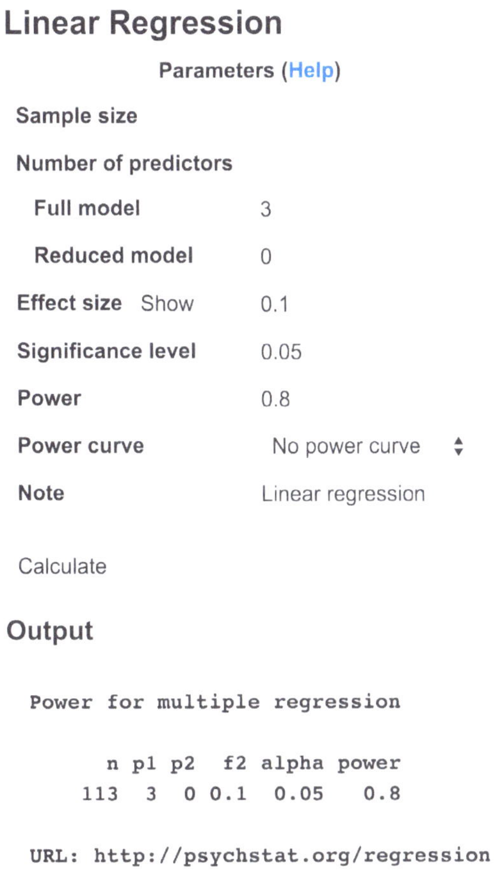

Linear Regression

Figure 9.1.5: Input and output for calculating sample size for regression in Example 9.1.3

Parameters (Help)

Sample size

Number of predictors

    Full model   3

    Reduced model   0

Effect size Show   0.1

Significance level   0.05

Power   0.8

Power curve   No power curve ⬍

Note   Linear regression

Calculate

Output

```
Power for multiple regression

    n p1 p2  f2 alpha power
  113  3  0 0.1  0.05   0.8


URL: http://psychstat.org/regression
```

```
   300  3  0 0.1  0.05 0.9981375

WebPower URL: http://psychstat.org/regression

> plot(res)  ## generate power curve
>
> ## sample size given effect size, sample size, power
> wp.regression(n=NULL, p1=3, f2=.1, power=0.8)

Multiple regression power calculation

         n p1 p2  f2 alpha power
  113.0103  3  0 0.1  0.05   0.8

WebPower URL: http://psychstat.org/regression

>
> ## calculate power given sample size and effect size and
    controling two predictors
> wp.regression(n=100, p1=3, p2=2, f2=.1429)
```

```
Multiple regression power calculation

     n p1 p2    f2 alpha      power
   100  3  2 0.1429  0.05 0.9594695

WebPower URL: http://psychstat.org/regression
```

9.3   *Effect Size for Linear Regression*

We use the effect size measure f^2 proposed by Cohen (1988, p.410) as the measure of the regression effect size. Cohen discussed the effect size in three different cases, which actually can be generalized using the idea of a full model and a reduced model as shown by Maxwell & Delaney (2003). The f^2 is defined as

$$f^2 = \frac{R^2_{Full} - R^2_{Reduced}}{1 - R^2_{Full}},$$

where R^2_{Full} and $R^2_{Reduced}$ are R-squared for the full and reduced models respectively. Cohen (1988) suggested that f^2 values of 0.02, 0.15, and 0.35 represent small, medium, and large effect sizes.

Suppose that a researcher wants to examine the effect of marital status, self-esteem, and social support on depression. From the literature, he concludes that the three variables can explain a small portion of the variance in depression ($R^2 = .0196$). What is the effect size in term of f^2?

This example is from the *Case 0* in Cohen (1988, p.409), which is the most common case in research involving regression. In this case, the reduced model does not contain any predictor, while the full model contains all three predictors under investigation. The R^2 of the full model is .0196, while the R^2 of the reduced model is 0. Therefore, the effect size f^2 is

$$f^2 = \frac{R^2_{Full} - R^2_{Reduced}}{1 - R^2_{F}ull} = \frac{0.0196 - 0}{1 - 0.0196} = .0199 \approx 0.02.$$

We can also use WebPower to calculate f^2 for this example. In the interface in Figure 9.1.1, first, we clicked on "**Show**" button after the *Effect Size* field, to expand the effect size calculation window, which is shown in Figure 9.3.1. We typed 0.0196 in the *R-squared* field in the *Full model* box. We left the *R-squared* in the *Reduced Model* to be zero. After clicking the "**Calculate**" button, the effect size calculation window disappeared and the effect size, 0.02, was filled in the *Effect size* field as in Figure 9.3.1.

Example 9.3.1: Effect size of overall test of regression coefficients

Example 9.3.2: Calculating the effect size for a subset of predictors in regression

Linear Regression

Figure 9.3.1: Input for calculating the effect size for regression in Example 9.3.1

Parameters (Help)

Sample size	100
Number of predictors	
Full model	3
Reduced model	2
Effect size Show	0.0200

Effect size calculation

R-squared for full model	0.0196
R-squared for reduced model	0
	Calculate

Significance level	0.05
Power	
Power curve	No power curve ↕
Note	Linear regression

Calculate

A researcher is planning to study the effect of parental involvement on students' achievement. She realizes that some demographic variables also correlate with students' achievement. Therefore, she wants to control these variables out of the effect of parental involvement on students' academic achievement. She chooses two variables that have shown significant relationship with students' achievement from previous studies: family income and parents' education. She expects the effect of parental involvement together with the two variables on students' achievement to be medium ($R^2 = .16$), while the effect of the two demographic variables alone on students' achievement to be small ($R^2 = .04$). What is the effect size of parental involvement on students' achievement controlling for family income and parents' education?

This problem is from the example in *Case 1* in Cohen (1988, p.409). In this case, the reduced model contains two control variables, family income, and parents' education, while the full model contains all three variables. Then $R^2_{Full} = .16$ and $R^2_{Reduced} = 0.04$. Therefore,

$$f^2 = \frac{R^2_{Full} - R^2_{Reduced}}{1 - R^2_{Full}} = \frac{0.16 - 0.04}{1 - 0.16} = .143.$$

WebPower can be similarly used to get the effect size in this example.

9.4 *Technical Details*

Let y_i denote the measurement of a dependent variable for the ith individual, x_{ij} denote the measurement of the jth independent variable for the ith individual, and β_j denote the coefficient of the effect of the jth independent variable on the dependent variable. The regression model can be expressed as follows:

$$y_i = \beta_0 + \sum_{j=1}^{p} \beta_j x_{ij} + e_i$$

or in the matrix form as:

$$\mathbf{y} = \mathbf{X}\beta + \mathbf{e}$$

where $\mathbf{y}$ denotes an $n \times 1$ vector of the dependent variable for n individuals, $\mathbf{X}$ denotes a $n \times (p+1)$ matrix of p independent variables and a column of 1, and $\mathbf{e}$ denotes an $n \times 1$ vector of residuals.

The hypothesis test in regression can be considered as a tool for choosing between a reduced model and the associated full model (Rencher & Schaalje, 2008). In the case of the omnibus test, the null hypothesis states that all parameters except for the intercept are zero:

$$H_0 : \beta_1 = \beta_2 = \cdots = \beta_p = 0.$$

The alternative hypothesis states that at least one of the parameters is not equal to zero:

$$H_1 : \exists j; \beta_j \neq 0, \quad j = 1, 2, 3, \ldots, p.$$

Estimating the model gives

$$\hat{\beta} = (\mathbf{X}'\mathbf{X})^{-1}\mathbf{X}'\mathbf{y},$$
$$\hat{\sigma}^2 = (\mathbf{y}-\mathbf{X}\hat{\beta})'(\mathbf{y}-\mathbf{X}\hat{\beta})/n - p - 1.$$

An F statistic can be obtained to test $H_0 : \beta_1 = \beta_2 = \cdots = \beta_p = 0$,

$$F = \frac{SSR/p}{SSE/(n-p-1)}$$

where

$$SSR = \hat{\beta}'\mathbf{X}'\mathbf{y} - n\bar{y}^2,$$
$$SSE = \mathbf{y}'\mathbf{y} - \hat{\beta}'\mathbf{X}'\mathbf{y}.$$

The F statistic follows an F distribution with degrees of freedom $u = p$ and $v = n - p - 1$ if the null hypothesis is true. If it is larger than the critical value $F_{1-\alpha}$, one would reject the null hypothesis.

Under the alternative hypothesis, the statistic follows a non-central F distribution with the non-central parameter (Cohen, 1988, p.414):

$$\lambda = (u + v + 1)f^2 = nf^2.$$

More generally, we can test whether a set of regression coefficients are equal to 0. Then the F statistic is

$$F = \frac{(SSE_{Reduced} - SSE_{Full})/(df_{E_{Reduced}} - df_{E_{Full}})}{(SSE_{Full}/df_{E_{Full}})} \qquad (9.4.1)$$

and the non-central parameter is

$$\lambda = (u + v + 1)f^2 = (df_{E_{Full}} + df_{E_{Reduced}} + 1)f^2$$

where $v = df_{E_{Full}} = n - p - 1$ and $u = df_{E_{Reduced}} = p - p^*$ with p^* denoting the number of control variables or the number of variables in the reduced model. With the non-central parameter, the power is obtained as

$$\pi = 1 - F_{u,v,\lambda}(F_{u,v,1-\alpha}^{-1}).$$

9.5   Exercises

1. A study will be conducted to investigate the effect of internal motivation on student achievement controlling for intelligence, family income, and parental involvement. From the literature review, the effect of internal motivation together with the controlling variables, are expected to be medium to high (R^2=0.5), while the effect of the three controlling variables together on achievement is small (R^2=0.14). How many participants should this study recruit if we want to have a power 0.8 to reject the null hypothesis with the α level 0.05? Make a power curve to show the power of the analysis involving sample size from 100 to 1000 (you can choose your own interval).

2. A researcher plans to conduct a study examining the effect of sleeping hours on people's well being. She also includes 2 other covariates of psychological well being in the study: years of education and self-esteem. The effect of sleeping hours, together with the other two variables, on well being is expected to be moderate (R^2=.39), while from the literature it is known that the effect of years in education and self-esteem together on psychological well being is also medium (R^2=0.3). Due to budget restriction, she can only recruit 50 participants for her study. Please help her calculate the power for her study. Should she add more participants in your opinion? Why?

10 Statistical Power Analysis for Logistic Regression

Haiyan Liu
Psychological Sciences
University of California, Merced

Logistic regression is a type of generalized linear models widely used to model the association between a binary outcome variable and predictors. Maximum likelihood estimation methods can be used to obtain model parameter estimates of a logistic regression model. To test the significance of the effect of a predictor, the Wald test is commonly used. A Wald test is a statistical hypothesis test in which a vector of parameter estimates is compared against the expected means under the null hypothesis and weighted by the precision matrix. The power analysis method in Demidenko (2007) is used in WebPower. To compute the standard errors of parameter estimates for power analysis, the distributions of the predictors are required under the population level, for which the WebPower provides options including the widely used distributions such as Bernoulli, exponential, lognormal, normal, Poisson, and uniform distributions.

10.1 How to Conduct Power Analysis for Logistic Regression

Power analysis for logistic regression can be conducted by using the online software WebPower with the interface shown in Figure 10.1.1. To conduct a power analysis, one needs to specify $Prob(Y = 1 | X = 0)$ – the probability of $Y = 1$ when $X = 0$, $Prob(Y = 1 | X = 1)$ – the probability of $Y = 1$ when $X = 1$, and the distribution of the predictor, which are used to compute the population regression coefficients and standard errors of the estimated regression coefficients. Through this interface, one can compute either *Sample size*, *Power*, or *Significance level* with any two of them known.

http://psychstat.org/logistic

- *Sample size* represents the number of observations in the study. To

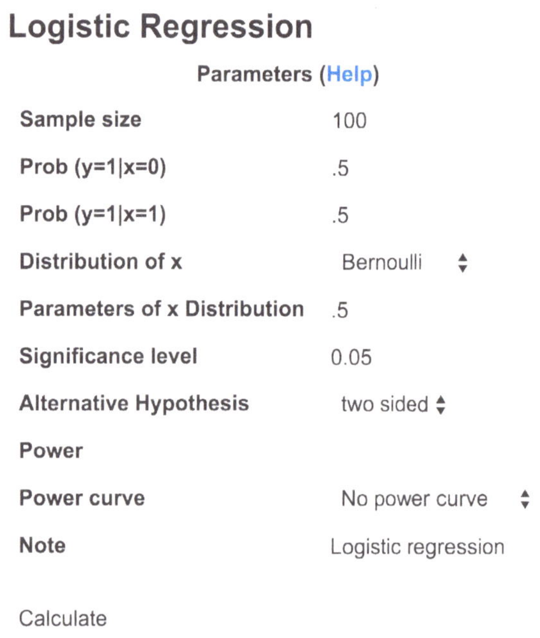

Logistic Regression

Parameters (Help)

Sample size	100
Prob (y=1|x=0)	.5
Prob (y=1|x=1)	.5
Distribution of x	Bernoulli ⬍
Parameters of x Distribution	.5
Significance level	0.05
Alternative Hypothesis	two sided ⬍
Power	
Power curve	No power curve ⬍
Note	Logistic regression

Calculate

Figure 10.1.1: Software interface of power analysis for simple logistic regression modeling

compute power, sample size needs to be specified before the analysis. There are two ways to input the sample size. One is to input one sample size each time and compute a power. The other is to input multiple sample sizes, and output multiple power values simultaneously. To input multiple sample sizes, users can either separate the multiple sample sizes by white spaces (e.g., 100 150 200) or use the method s:e:i with s denoting the starting sample size, e as ending sample size, and i as the interval. For example, 100:150:10 will generate a sequence 100 110 120 130 140 150. A default sample size 100 is filled in the interface as shown in Figure 10.1.1, which can be replaced by users.

- *Prob*$(Y = 1|X = 0)$ is the probability of observing 1 for the outcome variable when the predictor $X = 0$, from which the intercept parameter can be computed.

- *Prob*$(Y = 1|X = 1)$ is the probability of observing 1 for the outcome variable when the predictor $X = 1$. The regression coefficient of the predictor X can be computed from *Prob*$(Y = 1|X = 0)$ and *Prob*$(Y = 1|X = 1)$.

- *Distribution of X* specifies the population distribution of the predictor. Six options are provided: Bernoulli(π), Exponential(λ), lognormal(μ, σ), normal(μ, σ), Poisson(λ), and uniform$[L, R]$. Through *Parameter of X Distribution*, one can input the values of the parameters of the distribution of the predictor. For the distributions with multiple parameters, for example, lognormal, normal, and uniform distributions, the values of parameters are specified in the order as they appear in the usual way and separated by white spaces. For instance, for the *Parameter of X Distribution*, 0.5 1.2 are filled in and separated by a white space. It means $\mu = 0.2, \sigma = 1.2$ if the predictor follows a lognormal/normal distribution, and $L = 0.5, R = 1.2$ if the predictor follows a uniform distribution.

- The *Significance level* (Type I error rate) for power calculation is needed (default 0.05).

- *Alternative Hypothesis* tells the type of the alternative hypothesis. It is either "Two-sided" (default), "greater" or "less".

- The *Power* specifies the statistical power of the analysis. It is left empty if the power is going to be computed in the analysis (then the sample size must be specified), or is specified a priori number (e.g., 0.80) to compute the sample size.

- *Power curve* can be plotted if multiple sample sizes are provided and "*Show power curve*" is set.

10.1.1   *Examples*

Researchers are interested in the association between GPA, which is assumed to follow a normal distribution $N(0,1)$ after standardization, and marijuana use, whether a participant has used marijuana or not before. For this study, the data could be analyzed by logistic regression for its outcome variable is binary. If they already know that in the population, students with standardized GPA scores 0 and 1 are expected to use marijuana with probabilities 0.15 and 0.10, respectively. Then what is the power to detect the effect of GPA on marijuana use in a study with 200 participants?

Example 10.1.1: Compute statistical power given the sample size

The input and output for calculating the power for this study are given in Figure 10.1.2. In the field of *Sample size*, input 200 and in the field of $Prob(Y = 1|X = 0)$, input 0.15, and in the field of $Prob(Y = 1|X = 1)$, input 0.10. The default significance level 0.05 is used. In the field of *Distribution of X*, choose "Normal" and the in the field "*Parameters of X Distribution*", input 0 1, indicating the mean of the normal distribution is 0 and the standard error is 1. Since we need to calculate power, the field for *Power* is left blank. By clicking the "**Calculate**" button, the statistical power is given as 0.6299 in the output.

Example 10.1.2: Compute the sample size given power

In this example, researchers are planning a study on the association between GPA and marijuana use. They hope their study to have a

Logistic Regression

Parameters (Help)

Sample size	200
Prob (y=1|x=0)	0.15
Prob (y=1|x=1)	0.1
Distribution of x	Normal ▲▼
Parameters of x Distribution	0 1
Significance level	0.05
Alternative Hypothesis	two sided ▲▼
Power	
Power curve	No power curve ▲▼
Note	Logistic regression

Calculate

Output

```
Power for logistic regression

    p0  p1  beta0   beta1   n alpha  power
  0.15 0.1 -1.735 -0.4626 200  0.05 0.6299

URL: http://psychstat.org/logistic
```

Figure 10.1.2: Input and output for calculating power for logistic regression with sample size 200 in Example 10.1.1

power at least 0.8. Then how many participants should they recruit?

The input and output for this analysis are given in Figure 10.1.3. Leave the filed *Sample size* blank, and in the field *"Power"*, input 0.8 (the desired power for the study). By clicking the "**Calculate**" button, the sample size is given at the bottom. For this study, the sample size is 298.9. Thus, at least 299 participants are needed for the study to reach a power 0.80.

In the case a researcher wants to explore the effect of sample size on the desired power, he/she can generate a power curve according to given sample sizes. The input and output for plotting a power curve are given in Figure 10.1.4. In the *Sample Size* field, the input is 100:400:50, indicating the smallest and largest sample sizes are 100 and 400. In the output, the power for the sample sizes 100, 150, 200, $\cdots$, and 400 is listed and the power curve is displayed at the bottom of the output as shown in Figure 10.1.5. From the power curve, one can infer that to reach power 0.80, the required sample size would be around 300.

Example 10.1.3: Generate a power curve for logistic regression

Logistic Regression

Parameters (Help)

Sample size

Prob (y=1|x=0)   0.15

Prob (y=1|x=1)   0.1

Distribution of x   Normal ⇕

Parameters of x Distribution   0 1

Significance level   0.05

Alternative Hypothesis   two sided ⇕

Power   0.8

Power curve   No power curve ⇕

Note   Logistic regression

Calculate

Output

```
Power for logistic regression

    p0  p1  beta0   beta1     n alpha power
  0.15 0.1 -1.735 -0.4626 298.9  0.05   0.8


URL: http://psychstat.org/logistic
```

10.2 Using R Package WebPower for Power Analysis for Logistic Regression

The power calculation for logistic regression is conducted using the R function wp.logistic. The detail of the function is:

```
wp.logistic(n = NULL, p0 = NULL, p1=NULL, alpha = 0.05, power =
    NULL, alternative = c("two.sided", "less", "greater"), family
    =c("Bernoulli", "exponential", "lognormal", "normal", "
    Poisson", "uniform"), parameter=NULL)
```

The R input and output for the examples used in this chapter are given below:

```
> ## calculate power given sample size and effect size
> wp.logistic(n=200, p0=.15, p1=.1,family="normal", parameter=c
    (0,1))

Power calculation for logistic regression
```

n: sample size
p0: $Prob(Y = 1|X = 0)$
p1: $Prob(Y = 1|X = 1)$
alpha: significance level
power: statistical power
alternative: alternative hypothesis
family: distribution of the predictor.
parameter: corresponding parameter for the predictor distribution. Default: Bernoulli: 0.5; exponential: 1; lognormal and normal: 0, 1; Poisson: 1; uniform: 0, 1.

Logistic Regression

Figure 10.1.4: Input and output for plotting a power curve of a simple logistic regression of Example 10.1.3

Parameters (Help)

Sample size	100:400:50
Prob (y=1|x=0)	0.15
Prob (y=1|x=1)	0.1
Distribution of x	Normal ⬍
Parameters of x Distribution	0 1
Significance level	0.05
Alternative Hypothesis	two sided ⬍
Power	
Power curve	Show power curve ⬍
Note	Logistic regression

Calculate

Output

```
Power for logistic regression

     p0  p1  beta0    beta1    n  alpha  power
    0.15 0.1 -1.735 -0.4626 100   0.05 0.3673
    0.15 0.1 -1.735 -0.4626 150   0.05 0.5099
    0.15 0.1 -1.735 -0.4626 200   0.05 0.6299
    0.15 0.1 -1.735 -0.4626 250   0.05 0.7265
    0.15 0.1 -1.735 -0.4626 300   0.05 0.8014
    0.15 0.1 -1.735 -0.4626 350   0.05 0.8580
    0.15 0.1 -1.735 -0.4626 400   0.05 0.8999


 URL: http://psychstat.org/logistic
```

```
    p0  p1    beta0      beta1   n  alpha    power
  0.15 0.1 -1.734601 -0.4626235 200 0.0125 0.6299315


WebPower URL: http://psychstat.org/logistic


>
> ## sample size given effect size, sample size, power
> wp.logistic(n=NULL, p0=.15, p1=.1,family="normal", parameter=c
    (0,1), power=0.8)


Power calculation for logistic regression

     p0  p1    beta0      beta1          n  alpha power
```

Figure 10.1.5: Power curve of a simple logistic regression of Example 10.1.3

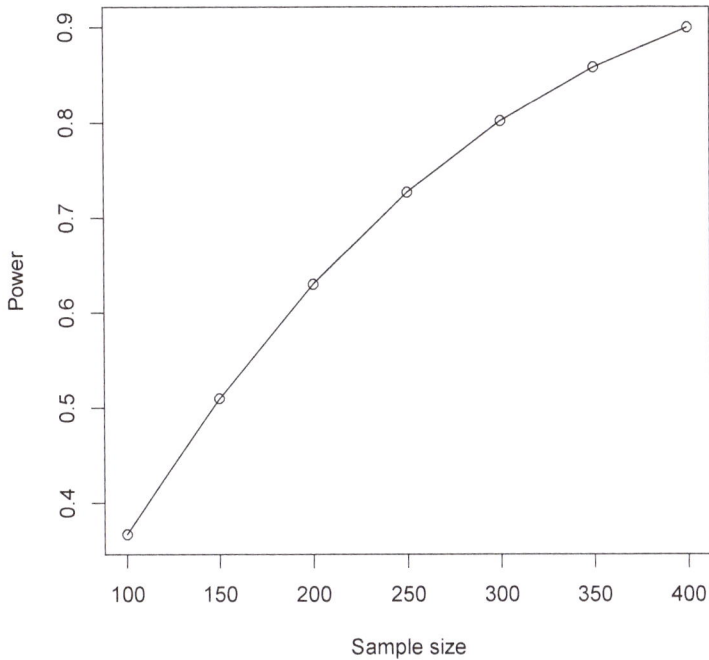

Figure 10.1.5: Power curve of a simple logistic regression of Example 10.1.3

```
    0.15 0.1 -1.734601 -0.4626235 298.9207 0.0125   0.8

WebPower URL: http://psychstat.org/logistic


>
> ## power curve
> res <- wp.logistic(n=seq(100, 500, 50), p0=.15, p1=.1,family="
    normal", parameter=c(0,1))
> res

Power calculation for logistic regression

    p0  p1    beta0      beta1  n alpha      power
   0.15 0.1 -1.734601 -0.4626235 100 0.0125 0.3672683
   0.15 0.1 -1.734601 -0.4626235 150 0.0125 0.5098635
   0.15 0.1 -1.734601 -0.4626235 200 0.0125 0.6299315
   0.15 0.1 -1.734601 -0.4626235 250 0.0125 0.7264597
   0.15 0.1 -1.734601 -0.4626235 300 0.0125 0.8014116
   0.15 0.1 -1.734601 -0.4626235 350 0.0125 0.8580388
   0.15 0.1 -1.734601 -0.4626235 400 0.0125 0.8998785
   0.15 0.1 -1.734601 -0.4626235 450 0.0125 0.9302222
   0.15 0.1 -1.734601 -0.4626235 500 0.0125 0.9518824


WebPower URL: http://psychstat.org/logistic
```

```
> plot(res)  ## generate power curve
```

<table>
<tr><td>**10.3**</td><td>*Technical Details*</td></tr>
</table>

The purpose of a logistic regression analysis is to model the conditional probability of observing "1" given covariates X. For the simple logistic regression, there is only one predictor, and

$$\Pr(Y = 1|X) = \frac{\exp(\beta_0 + \beta_1 X)}{1 + \exp(\beta_0 + \beta_1 X)}.$$

Let $\beta = (\beta_0, \beta_1)'$ represent the column vector of parameters and $\hat{\beta}_{ML} = (\hat{\beta}_0, \hat{\beta}_1)'$ be its maximum likelihood (ML) estimates. Under some regularity conditions, $\hat{\beta}_{ML}$ is normally distributed,

$$\sqrt{n}(\hat{\beta}_{ML} - \beta) \rightsquigarrow \text{MVN}(\mathbf{0}, I^{-1}),$$

where I is the expected Fisher-information matrix and n is the sample size. Let $l(\beta)$ represent the log-likelihood function, and according to the definition of Fisher-information matrix, we have

$$I = -E \begin{bmatrix} \frac{\partial^2 l(\beta)}{\partial \beta_0^2}) & \frac{\partial^2 l(\beta)}{\partial \beta_0 \beta_1}) \\ \frac{\partial^2 l(\beta)}{\partial \beta_0 \beta_1}) & \frac{\partial^2 l(\beta)}{\partial \beta_1^2}) \end{bmatrix} = \begin{bmatrix} I_{00} & I_{01} \\ I_{01} & I_{11} \end{bmatrix},$$

whose inverse matrix is $I^{-1} = \frac{1}{I_{00}I_{11} - I_{01}^2} \begin{bmatrix} I_{11} & -I_{01} \\ -I_{01} & I_{00} \end{bmatrix}$. As a consequence, the estimated variance of $\hat{\beta}_1$ with n observations is $\frac{1}{n} \frac{I_{00}}{I_{00}I_{11} - I_{01}^2}$ evaluated at $(\hat{\beta}_0, \hat{\beta}_1)$.

To test the significance of the effect of X, we need to conduct the following hypothesis testing,

$$H_0 : \beta_1 = 0$$

with the alternative hypothesis

$$H_a : \beta_1 \neq 0 \text{ (two sided)}.$$

In a Wald test, the variance of the parameter estimates is evaluated at the parameter estimates under the alternative hypothesis. The Wald test statistic

$$z_w = \frac{\hat{\beta}_1 - \beta_1}{\sqrt{var(\hat{\beta}_1)|_{(\hat{\beta}_0, \hat{\beta}_1)}}}$$

converges to the standard normal distribution.

The statistical power of a hypothesis testing is defined as the probability that the null hypothesis is rejected when the alternative hypothesis is true. To compute the power, we adopted the procedure introduced by Demidenko (2007). They computed power under the population level by assuming the population value of β_0 and β_1 as well as the distribution for predictors are known. According to the definition of statistical power, we have

$$
\begin{aligned}
\text{power} \quad &= \quad \Pr(H_0 \text{ is rejected} | H_a \text{ is true}) \\
&\stackrel{\text{def}}{=} \quad \Pr(|\frac{\hat\beta_1 - \beta_1 | H_0}{\sqrt{var(\hat\beta_1)|_{(\hat\beta_0,\hat\beta_1)}}}| > Z_{1-\frac{\alpha}{2}} | H_a \text{ is true}).
\end{aligned}
$$

For given population values (β_0, β_1), because $(\hat\beta_0, \hat\beta_1)$ are ML estimates, and by Slusky Theorem,

$$
var(\hat\beta_1)|_{(\hat\beta_0,\hat\beta_1)} \to var(\hat\beta_1)|_{(\beta_0,\beta_1)} \text{ as } n \text{ goes to infinity.}
$$

As a consequence,

$$
\begin{aligned}
\text{power} \quad &\approx \quad \Pr(|\frac{\hat\beta_1 - \beta_1 | H_0}{\sqrt{var(\hat\beta_1)|_{(\beta_0,\beta_1)}}}| > Z_{1-\frac{\alpha}{2}} | H_a \text{ is true}) \\
&= \quad \Pr(\frac{\hat\beta_1 - \beta_1 | H_0}{\sqrt{var(\hat\beta_1)|_{(\beta_0,\beta_1)}}} > Z_{1-\frac{\alpha}{2}} | H_a \text{ is true}) \\
&\quad + \Pr(\frac{\hat\beta_1 - \beta_1 | H_0}{\sqrt{var(\hat\beta_1)|_{(\beta_0,\beta_1)}}} < -Z_{1-\frac{\alpha}{2}} | H_a \text{ is true}).
\end{aligned}
$$

Because $\hat\beta_1$ is the ML estimate, $\hat\beta_1$ is normally distributed with mean β_1 and variance $var(\hat\beta_1)|_{(\beta_0,\beta_1)}$. Under the alternative hypothesis, $\beta_1 \neq 0$,

$$
\begin{aligned}
\text{power} \quad &\approx \quad \Pr(\frac{\hat\beta_1 - \beta_1 | H_1}{\sqrt{var(\hat\beta_1)|_{(\beta_0,\beta_1)}}} > Z_{1-\frac{\alpha}{2}} - \frac{\beta_1 | H_1 - \beta_1 | H_0}{\sqrt{var(\hat\beta_1)|_{(\beta_0,\beta_1)}}} | H_a \text{ is true}) \\
&\quad + \Pr(\frac{\hat\beta_1 - \beta_1}{\sqrt{var(\hat\beta_1)|_{(\beta_0,\beta_1)}}} < -Z_{1-\frac{\alpha}{2}} - \frac{\beta_1 | H_1 - \beta_1 | H_0}{\sqrt{var(\hat\beta_1)|_{(\beta_0,\beta_1)}}} | H_a \text{ is true}) \\
&\approx \quad \Phi(-Z_{1-\frac{\alpha}{2}} + \frac{\beta_1 | H_1 - \beta_1 | H_0}{\sqrt{var(\hat\beta_1)|_{(\beta_0,\beta_1)}}}) \quad\quad\quad (10.3.1) \\
&\quad + \Phi(-Z_{1-\frac{\alpha}{2}} - \frac{\beta_1 | H_1 - \beta_1 | H_0}{\sqrt{var(\hat\beta_1)|_{(\beta_0,\beta_1)}}}) \quad\quad\quad (10.3.2)
\end{aligned}
$$

where $var(\hat\beta_1)|_{(\beta_0,\beta_1)} = \frac{1}{n} \frac{I_{00}}{I_{00}I_{11} - I_{01}^2}|_{(\beta_0,\beta_1)}$, computed from the Fisher-information matrix for a given population model.

In the power expression in Equation (10.3.1), the power is related to the following factors:

1. The sample size n.

2. The significance level α.

3. (β_0, β_1) in the population: they could be obtained from $\Pr(Y = 1|X = 0)$ and $\Pr(Y = 1|X = 1)$, with

$$\beta_0 = \log \frac{\Pr(Y = 1|X = 0)}{1 - \Pr(Y = 1|X = 0)}$$

$$\beta_1 = \log \frac{\Pr(Y = 1|X = 1)/(1 - \Pr(Y = 1|X = 1))}{\Pr(Y = 1|X = 0)/(1 - \Pr(Y = 1|X = 0))}.$$

4. Fisher-information matrix I: obtained by taking the expectation of both Y and X. As a consequence, the users need to specify the distribution of the predictor X.

10.4   Exercises

1. A student believes that the number of friends is related to whether a person smokes cigars or not. To test the theory, she plans to collect data from 100 people to ask whether they smoke and how many friends they have. Suppose after standardizing the data on the number of friends, she estimates that the probability for a person to smoke is 0.05 when the standardized number of friends is 0 and the probability is 0.15 at the one standard deviation of the number of friends. Based on the information, what is the power to detect the relationship between smoking and the number of friends?

2. Using the information in Exercise 1, generate a power curve with the sample size ranging from 100 to 300. What is the required sample size to get a power at least 0.8 and 0.9?

11   *Statistical Power Analysis for Poisson Regression*

Haiyan Liu
Psychological Sciences
University of California, Merced

Poisson regression belongs to the family of generalized linear models and is used to model the count data. It assumes that the outcome variable, which only takes non-negative integer values, follows a Poisson distribution. The mean of a Poisson distribution is modeled as a linear combination of predictors. To estimate the model, maximum likelihood based methods can be used, from which both the parameter estimates and their standard error estimates are obtained. To test the significance of parameters, the Wald test is commonly used, which assumes the estimated regression coefficient follows a normal distribution with respective means under the null and alternative hypotheses. The standard error of the estimated regression coefficient is a function of both parameter estimates and predictors. In WebPower, we compute the standard errors of parameter estimates under the population level of predictors for power analysis. The users have a wide range of options for the distribution of the predictors including Bernoulli, exponential, lognormal, normal, Poisson, and uniform distributions.

11.1   *How to Conduct Power Analysis for Poisson Regression*

The Poisson regression model with one predictor is,

$$\Pr(Y = y | \lambda, X) \quad = \quad \frac{\lambda^y \exp(-\lambda)}{y!} \qquad (11.1.1)$$

$$\log \lambda \quad = \quad \beta_0 + \beta_1 X \qquad (11.1.2)$$

where λ is a parameter representing the average number of events/outcomes. The primary interest of Poisson regression is to test whether the predictor X is related to Y or not. In terms of hypothesis testing, the null and alternative hypotheses are

$$H_0 : \beta_1 = 0$$
$$H_1 : \beta_1 \neq 0.$$

To conduct the hypothesis test, the Wald statistics, which is asymptotically normally distributed, can be used:

$$W = \frac{\hat{\beta}_1}{se(\hat{\beta}_1)}$$

where $\hat{\beta}_1$ is the estimate of β_1 and $se(\hat{\beta}_1)$ is the standard error estimate of $\hat{\beta}_1$.

Power analysis for the simple Poisson regression can be conducted using the online software WebPower with the interface shown in Figure 11.1.1. In order to conduct power analysis, one needs to provide two quantities: Exp0 and Exp1. For given regression coefficients, Exp0 takes the value of $\exp(\beta_0)$, which represents the event rate under the null hypothesis; and Exp1 is $\exp(\beta_1)$, the ratio of $\exp[\beta_0 + \beta_1(x + 1)]$ and $\exp[\beta_0 + \beta_1 x]$, and it is the relative increase of the event rate corresponding to one unit change in X under the alternative hypothesis. Both Exp0 and Exp1 should be positive. Through the interface of WebPower, one can compute *Sample size*, *Power*, or *Significance level* given the rest of the information known.

http://psychstat.org/poisson

Poisson Regression

Parameters (Help)

Sample size	100
Exp0=exp(beta0): Base rate	.5
Exp1=exp(beta1): Rate change	.5
Distribution of x	Bernoulli ⬍
Parameters of x Distribution	.5
Significance level	0.05
Alternative Hypothesis	two sided ⬍
Power	
Power curve	No power curve ⬍
Note	Poisson regression

Calculate

Figure 11.1.1: Software interface of power analysis for Poisson regression

- *Sample size* represents the number of observations in the study. To compute power, sample size needs to be specified by users prior to the analysis. There are two ways to input sample sizes. One is to input one sample size and compute the power each time. The other is to input multiple sample sizes, and output multiple power values simultaneously. To input multiple sample sizes, users can either separate multiple sample sizes by white spaces (e.g., `100 150 200`) or use the method `s:e:i` with `s` denoting the starting sample size, `e` as the ending sample size, and `i` as the interval. For example, `100:150:10` will generate a sequence of values: `100 110 120 130 140 150`.

- *Exp0* is the base rate $\exp(\beta_0)$ under the null hypothesis, which always takes a positive value.

- *Exp1* is the relative increase of the event rate, $\exp(\beta_1)$, with respect to one unit increase on X under the alternative hypothesis, which is used to compute the effect size.

- *Distribution of X* specifies the population distribution of the predictor. Six options are provided: Bernoulli(π), Exponential(λ), lognormal(μ, σ), normal(μ, σ), Poisson(λ), and uniform$[L, R]$.

- Through *Parameter of X Distribution*, one can input the values of the parameters of the distribution of the predictor. For the distributions with multiple parameters, for example, lognormal, normal, and uniform distributions, the values of parameters are specified in the order as they appear in the usual way and separated by white spaces. For instance, for the *Parameter of X Distribution*, if 0.5 1.2 are filled in and separated by a white space, it means $\mu = 0.2, \sigma = 1.2$ if the predictor follows a lognormal/normal distribution, and $L = 0.5, R = 1.2$ if the predictor follows a uniform distribution.

- The *Significance level* (Type I error rate) for power calculation is needed (default 0.05).

- *Alternative Hypothesis* tells the type of the alternative hypothesis. It is either "Two-sided" (default), "greater" or "less".

- The *Power* specifies the statistical power of the analysis. It is left empty if the power is going to be computed in the analysis (then the sample size must be specified), or specified a priori number (e.g.,0.80) to compute the sample size.

- *Power curve* can be plotted if multiple sample sizes are provided and "*Show power curve*" is set.

11.1.1 *Examples*

Example 11.1.1: Compute power given the sample size

This example uses the results from Deb et al. (1997). In that study, researchers analyzed data on 4406 individuals, aged 66 and over, who were covered by Medicare. The purpose of that study was to model the relation between the demand of medical care, which was measured by the number of physician office visits, and the covariates related to the patients, for instance, gender of the patients. The outcome variable is the number of the physician office visits. Assume in the population Poisson regression model, the intercept and slope are 1.0289 and -0.1123 respectively. The predictor gender follows the Bernoulli distribution with mean 0.53. The question is, with a sample size 4406, what is the expected power?

The input and output for calculating the power for this study are given in Figure 11.1.2. In the field of *Sample size*, input 4406 and in the field of Exp0 input 2.798 , which is computed from $Exp0 = \exp(\beta_0)$. And in the field of Exp1, input 0.8938, which is computed by $Exp1 = \exp(\beta_1)$. In the field of *Distribution of X*, choose "Bernoulli", and in the field *Parameters of X Distribution*, input 0.53, indicating the mean of the Bernoulli distribution is 0.53. The default significance level 0.05 is used. A two-sided test is conducted by setting the *Alternative Hypothesis* to be "Two-sided". Since the power needs to be calculated, the field for *Power* is left blank. By clicking the "**Calculate**" button, the statistical power is given at the bottom, which is 1 for this example.

Example 11.1.2: Compute the sample size given the desired power

In Example 11.1.1, the power is 1. Suppose a researcher wants to have a power 0.8 with other information the same as in the previous example, how many participants does she need to recruit?

The input and output for this study are given in Figure 11.1.3. Note that the field *Sample size* is left blank and the power is set at 0.8. Based on the output, to reach the power 0.8, a sample size 944 is needed.

Example 11.1.3: Power curve for Poisson regression

Suppose in the example above, the researchers would like to know how the power changes as the sample size increases, e.g., from 600 to 1400. Then, a power curve can be generated. The input and output for plotting a power curve are given in Figure 11.1.4. In the *Sample Size* field, the input is 600:1400:200, indicating the smallest and largest sample sizes are 600 and 1400, respectively. In the output, the power for the sample sizes 600, 800, ..., 1400 is listed as shown in Figure 11.1.4 and the power curve is displayed at the bottom of the output as shown in Figure 11.1.5. From the power curve, one can infer that to reach a power 0.80, the sample size must be around 950, and to reach a power 0.9, the sample size needs to be about 1300. The power curve offers information on the relationship between the sample size and the power of a study visually.

Poisson Regression

Parameters (Help)

Sample size	4406
Exp0=exp(beta0): Base rate	2.798
Exp1=exp(beta1): Rate change	0.8938
Distribution of x	Bernoulli ⬍
Parameters of x Distribution	.53
Significance level	0.05
Alternative Hypothesis	two sided ⬍
Power	
Power curve	No power curve ⬍
Note	Poisson regression

Calculate

Output

```
Power for Poisson regression

    n power alpha  exp0   exp1 beta0   beta1 paremeter
 4406     1  0.05 2.798 0.8938 1.029 -0.1123      0.53


URL: http://psychstat.org/poisson
```

11.2   *Using R Package WebPower for Power Analysis for Poisson Regression*

The power calculation for Poisson regression is conducted using the R function wp.poisson. The detail of the function is:

```
wp.poisson(n = NULL, exp0 = NULL, exp1 = NULL, alpha = 0.05,
    power = NULL, alternative = c("two.sided", "less", "greater")
    , family = c("Bernoulli", "exponential", "lognormal", "normal
    ", "Poisson", "uniform"), parameter = NULL)
```

The R input and output for the examples in the chapter are given below:

```
> wp.poisson(n=4406, exp0=2.798, exp1=.8938, family='Bernoulli',
    parameter=.53)

Power calculation of Wald-test for poisson regression

    n    power alpha  exp0   exp1  beta0  beta1  paremeter
 4406    1    0.025 2.798 0.8938 1.0289 -0.1123      0.53
```

n: sample size
exp0: $\exp(\beta_0)$
exp1: $\exp(\beta_1)$
alpha: significance level
power: statistical power
alternative: alternative hypothesis
family: distribution of the predictor.
parameter: corresponding parameter for the predictor distribution. Default: Bernoulli: 0.5; exponential: 1; lognormal and normal: 0, 1; Poisson: 1; uniform: 0, 1.

Poisson Regression

Parameters (Help)

Sample size

Exp0=exp(beta0): Base rate	2.798
Exp1=exp(beta1): Rate change	0.8938
Distribution of x	Bernoulli ↕
Parameters of x Distribution	.53
Significance level	0.05
Alternative Hypothesis	two sided ↕
Power	0.8
Power curve	No power curve ↕
Note	Poisson regression

Calculate

Figure 11.1.3: Input and output for calculating sample size for Poisson regression with the desired power 0.80 in Example 11.1.2

Output

```
Power for Poisson regression

        n power alpha  exp0   exp1 beta0   beta1 paremeter
    943.3   0.8  0.05 2.798 0.8938 1.029 -0.1123      0.53


URL: http://psychstat.org/poisson
```

```
WebPower URL: http://psychstat.org/poisson

> wp.poisson(n=NULL, power=.8, exp0=2.798, exp1=.8938, family='Bernoulli
    ', parameter=.53)

Power calculation of Wald-test for poisson regression

        n power alpha  exp0   exp1     beta0    beta1 paremeter
    943.2   0.8 0.025 2.798 0.8938 1.028905  -0.1123      0.53

WebPower URL: http://psychstat.org/poisson

> res <- wp.poisson(n=seq(800, 1500, 100), exp0=2.798, exp1=.8938, family
    ='Bernoulli', parameter=.53)
> res

Power calculation of Wald-test for poisson regression

     n power alpha exp0 exp1 beta0 beta1 parameter
   800  0.73  0.05  2.8 0.89  1.03 -0.11      0.53
   900  0.78  0.05  2.8 0.89  1.03 -0.11      0.53
  1000  0.82  0.05  2.8 0.89  1.03 -0.11      0.53
  1100  0.86  0.05  2.8 0.89  1.03 -0.11      0.53
  1200  0.88  0.05  2.8 0.89  1.03 -0.11      0.53
```

Poisson Regression

Parameters (Help)

Sample size	600:1400:200
Exp0=exp(beta0): Base rate	2.798
Exp1=exp(beta1): Rate change	0.8938
Distribution of x	Bernoulli ⬍
Parameters of x Distribution	.53
Significance level	0.05
Alternative Hypothesis	two sided ⬍
Power	
Power curve	Show power curve ⬍
Note	Poisson regression

Calculate

Figure 11.1.4: Input and output for plotting a power curve for Poisson regression in Example 11.1.3

Output

```
Power for Poisson regression

     n  power alpha  exp0   exp1 beta0   beta1 paremeter
   600 0.6081  0.05 2.798 0.8938 1.029 -0.1123      0.53
   800 0.7324  0.05 2.798 0.8938 1.029 -0.1123      0.53
  1000 0.8224  0.05 2.798 0.8938 1.029 -0.1123      0.53
  1200 0.8849  0.05 2.798 0.8938 1.029 -0.1123      0.53
  1400 0.9269  0.05 2.798 0.8938 1.029 -0.1123      0.53


URL: http://psychstat.org/poisson
```

```
 1300  0.91  0.05  2.8 0.89  1.03 -0.11      0.53
 1400  0.93  0.05  2.8 0.89  1.03 -0.11      0.53
 1500  0.94  0.05  2.8 0.89  1.03 -0.11      0.53

WebPower URL: http://psychstat.org/poisson

> plot(res)  ## generate power curve
```

11.3   *Technical Details*

A simple Poisson regression model is

$$\Pr(Y_i = y_i | \lambda_i, X_i) = \frac{\lambda_i^{y_i} \exp(-\lambda_i)}{y_i!}$$
$$\log \lambda_i = \beta_0 + \beta_1 X_i.$$

Figure 11.1.5: Power curve for Poisson regression in Example 11.1.3

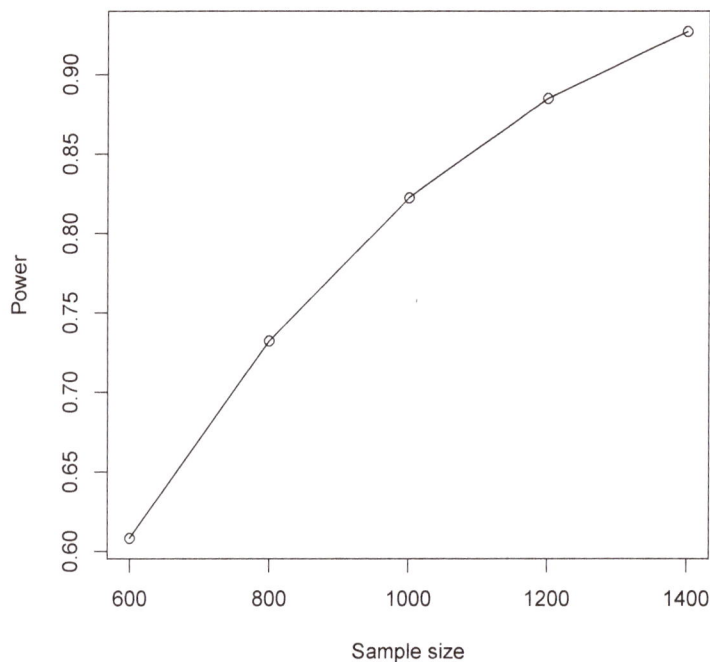

The log-likelihood function of the model is

$$l = \sum_{i=1}^{n} [y_i \log \lambda_i - \lambda_i - \log(y_i!)].$$

Let $\boldsymbol{\beta} = (\beta_0, \beta_1)'$ represent the column vector of parameters and $\hat{\boldsymbol{\beta}}_{ML} = (\hat{\beta}_0, \hat{\beta}_1)'$ be its maximum likelihood (ML) estimates. $\hat{\boldsymbol{\beta}}_{ML}$ follows a normal distribution with mean $\boldsymbol{\beta}$,

$$\sqrt{n}(\hat{\boldsymbol{\beta}}_{ML} - \boldsymbol{\beta}) \rightsquigarrow \text{BVN}(\mathbf{0}, \boldsymbol{I}^{-1}) \qquad (11.3.1)$$

where I is the expected Fisher-information matrix and n is the sample size. The information matrix can be obtained as

$$I = -E \begin{bmatrix} \frac{\partial^2 l(\beta)}{\partial \beta_0^2}) & \frac{\partial^2 l(\beta)}{\partial \beta_0 \beta_1}) \\ \frac{\partial^2 l(\beta)}{\partial \beta_0 \beta_1}) & \frac{\partial^2 l(\beta)}{\partial \beta_1^2}) \end{bmatrix} = \begin{bmatrix} I_{00} & I_{01} \\ I_{01} & I_{11} \end{bmatrix}$$

with

$$
\begin{aligned}
\frac{\partial l(\beta_0, \beta_1)}{\partial \beta_0} &= [\frac{y}{\lambda} - 1]\frac{\partial \lambda}{\partial \beta_0} = [y - \lambda] \\
\frac{\partial l(\beta_0, \beta_1)}{\partial \beta_1} &= [\frac{y}{\lambda} - 1]\frac{\partial \lambda}{\partial \beta_1} = [y - \lambda]x \\
I_{00} = \frac{\partial l^2(\beta_0, \beta_1)}{\partial \beta_0^2} &= -\exp(\beta_0 + \beta_1 x) \\
I_{11} = \frac{\partial l_i^2(\beta_0, \beta_1)}{\partial \beta_1^2} &= -x^2 \exp(\beta_0 + \beta_1 x_i) \\
I_{01} = \frac{\partial l^2(\beta_0, \beta_1)}{\partial \beta_0 \beta_1} &= -x \exp(\beta_0 + \beta_1 x_i).
\end{aligned}
$$

Given the sample size n, the estimated covariance matrix of $\hat{\boldsymbol{\beta}}_{ML}$ is $\frac{1}{n}I^{-1}$. Especially, the estimated variance of $\hat{\beta}_1$ is $\frac{1}{n}\frac{I_{00}}{I_{00}I_{11}-I_{01}^2}|_{(\hat{\beta}_0,\hat{\beta}_1)}$.

For the hypothesis testing with the null hypothesis

$$H_0 : \beta_1 = 0$$

and the alternative hypothesis

$$H_a : \beta_1 \neq 0 \text{ (two sided)},$$

the Wald test statistic is

$$z_w = \frac{\hat{\beta}_1 - \beta_1 | H_0}{\sqrt{var(\hat{\beta}_1)|_{(\hat{\beta}_0,\hat{\beta}_1)}}} \rightsquigarrow N(0, 1).$$

To compute the power, we adopted the procedure introduced by Demidenko (2007). They computed power based on the population value of β_0 and β_1 as well as the distribution of the predictors. According to the definition of statistical power, we have

$$
\begin{aligned}
\text{power} &\overset{\text{def}}{=} \Pr(H_0 \text{ is rejected}|H_a \text{ is true}). \\
&= \Pr(|\frac{\hat{\beta}_1 - \beta_1 | H_0}{\sqrt{var(\hat{\beta}_1)|_{(\hat{\beta}_0,\hat{\beta}_1)}}}| > Z_{1-\frac{\alpha}{2}}|H_a \text{ is true}).
\end{aligned}
$$

For given population values (β_0, β_1), because $(\hat{\beta}_0, \hat{\beta}_1)$ are the ML estimates, and by the Slusky Theorem,

$$var(\hat{\beta}_1)|_{(\hat{\beta}_0,\hat{\beta}_1)} \rightarrow var(\hat{\beta}_1)|_{(\beta_0,\beta_1)} \text{ as } n \text{ tends to infinity.}$$

Therefore,

$$
\begin{aligned}
\text{power} \quad &\approx \quad \Pr\!\left(\left| \frac{\hat{\beta}_1 - \beta_1 | H_0}{\sqrt{var(\hat{\beta}_1)|_{(\beta_0,\beta_1)}}} \right| > Z_{1-\frac{\alpha}{2}} \middle| H_a \text{ is true} \right) \\[2mm]
&= \quad \Pr\!\left(\frac{\hat{\beta}_1 - \beta_1 | H_0}{\sqrt{var(\hat{\beta}_1)|_{(\beta_0,\beta_1)}}} > Z_{1-\frac{\alpha}{2}} \middle| H_a \text{ is true} \right) \\[2mm]
&\quad + \Pr\!\left(\frac{\hat{\beta}_1 - \beta_1 | H_0}{\sqrt{var(\hat{\beta}_1)|_{(\beta_0,\beta_1)}}} < -Z_{1-\frac{\alpha}{2}} \middle| H_a \text{ is true} \right).
\end{aligned}
$$

Under the alternative hypothesis, $\beta_1 \neq 0$,

$$
\begin{aligned}
\text{power} \quad &\approx \quad \Pr\!\left(\frac{\hat{\beta}_1 - \beta_1 | H_1}{\sqrt{var(\hat{\beta}_1)|_{(\beta_0,\beta_1)}}} > Z_{1-\frac{\alpha}{2}} - \frac{\beta_1 | H_1 - \beta_1 | H_0}{\sqrt{var(\hat{\beta}_1)|_{(\beta_0,\beta_1)}}} \middle| H_a \text{ is true} \right) \\[2mm]
&\quad + \Pr\!\left(\frac{\hat{\beta}_1 - \beta_1 | H_1}{\sqrt{var(\hat{\beta}_1)|_{(\beta_0,\beta_1)}}} < -Z_{1-\frac{\alpha}{2}} - \frac{\beta_1 | H_1 - \beta_1 | H_0}{\sqrt{var(\hat{\beta}_1)|_{(\beta_0,\beta_1)}}} \middle| H_a \text{ is true} \right) \\[2mm]
&\approx \quad \Phi\!\left(-Z_{1-\frac{\alpha}{2}} + \frac{\beta_1 | H_1}{\sqrt{var(\hat{\beta}_1)|_{(\beta_0,\beta_1)}}} \right) + \Phi\!\left(-Z_{1-\frac{\alpha}{2}} - \frac{\beta_1 | H_1 - \beta_1 | H_0}{\sqrt{var(\hat{\beta}_1)|_{(\beta_0,\beta_1)}}} \right) \quad (11.3.2)
\end{aligned}
$$

where $var(\hat{\beta}_1)|_{(\beta_0,\beta_1)} = \frac{1}{n} \frac{I_{00}}{I_{00}I_{11}-I_{01}^2}|_{(\beta_0,\beta_1)}$, computed from the Fisher-information matrix for a given population model.

11.4   *Exercises*

1. A school is interested in the relationship between the days of absence in a year and the students' language test scores. A researcher plans to collect the academic information on 316 students. The response variable is the days of absence during a school year and the predictor is the standardized language test score, which is standardized to follow a normal distribution with mean 0 and standard deviation 1. A Poisson model will be used to analyze the data. Suppose at the population level, the intercept and slope are 2.30 and -0.12. What will be the power of this study?

2. With the same information as in the above exercise, what is the sample size needed to achieve a power 0.90?

12 *Statistical Power Analysis for Cluster Randomized Trials*

Miao Yang
Department of Psychology
University of Notre Dame

Multilevel designs for cross-sectional studies typically contain cluster randomized trials and multisite randomized trials (Liu, 2013). Cluster randomized trials (CRT) are used when the entire cluster is randomly assigned to either a treatment arm or a control arm. The data from CRT can be analyzed in a two-level hierarchical linear model, where the indicator variable for treatment assignment is included in the second level. If a study contains multiple treatments, then multiple indicators will be used. Our power analysis for CRT will focus on designs with 2 arms (i.e., a treatment and a control) and 3 arms (i.e., two treatments and a control).

12.1 *How to Conduct Power Analysis for CRT with 2 Arms*

The primary software interface for power analysis for CRT with two arms (i.e., one treatment and one control) is shown in Figure 12.1.1. Within the interface, a user can supply different parameter values and select different options for power analysis. Among the six parameters, *Sample size*, *Effect size*, *Number of clusters*, *Intra-class correlation*, *Significance level*, and *Power*, one and only one can be left blank.

http://psychstat.org/crt2arm

- The *Sample size* is the number of individuals within each cluster. The power calculation assumes a balanced design – equal sample size for each cluster. To obtain power estimation with varied sample sizes, multiple sample sizes can be provided in the following two ways. First, multiple sample sizes can be supplied and separated by white spaces, e.g., 100 150 200 will calculate power for the three sample sizes 100, 150 and 200. Second, a sequence of sample (or cluster) sizes can be generated using the method s:e:i with s denoting

the starting sample size, e as the ending sample size, and i as
the interval. Note that the values are separated by colon ":". For
example, 100:150:10 will generate a sequence of sample sizes: 100
110 120 130 140 and 150. By default, the sample size is 100.

- The *Effect size* specifies the main effect of treatment, the mean difference between the treatment clusters and the control clusters. Multiple
effect sizes or a sequence of effect sizes can be supplied using the
same way as for sample size. The default value is 0.6. One can either
input the effect size directly or calculate it by clicking the "**Calculator**"
link in the parentheses. Note that the calculator also allows users to
obtain intra-class correlation given the required information.

- The *Number of clusters* tells how many clusters are considered in the
study design. More than two clusters are required. Multiple cluster
sizes or a sequence of cluster sizes can be supplied using the same
way for sample size.

- The *Intra-class correlation* is the ratio of between-cluster variance to the
total variance, which quantifies the degree to which two randomly
drawn observations within a cluster are correlated. The default value
of intra-class correlation is 0.15. One can either input the intra-class
correlation directly or calculate it by clicking the "**Calculator**" link
after *Effect size*.

- The *Power* specifies the desired statistical power, usually set at 0.80.

- The *Type of analysis* can be specified as "Two-sided test" and "One-
sided test". The corresponding alternative hypotheses are:

$$"Two-sided" :\ \mu_D \neq 0,$$

$$"One-sided" :\ \mu_D > 0 \text{ or } \mu_D < 0,$$

where μ_D is the mean difference between the treatment clusters and
the control clusters.

- The *Significance level* for power calculation is needed but usually set
at the default value 0.05.

- In addition to the required input, one can also request the plot of a
power curve if multiple sample sizes, or cluster sizes, or effect sizes
are provided.

- A note (less than 200 characters) can also be provided to provide
basic information on the analysis for future reference for registered
users.

Once all fields have been filled in, pressing "**Calculate**" will create a table of output and, if requested, a power curve will appear below the table.

12.1.1   *Examples*

Cluster randomized trials with 2 arms

Figure 12.1.1: Software interface of power analysis for CRT with 2 arms

Parameters (Help)	
Sample size	100
Effect size (Calculator)	0.6
Number of clusters	16
Intra-class correlation	0.15
Power	
Significance level	0.05
Power curve	No power curve ⬍
H1	Two-sided test ⬍
Note	Cluster randomized tria

Calculate

Example 12.1.1: Calculate power given sample sizes, effect size and intra-class correlation

A group of education researchers developed a new teaching method to help students improve their math ability. They plan to randomly assign 5 classrooms to the new method and 5 classrooms to the standard method. Each classroom has 20 students. Based on their prior knowledge, they hypothesize that the effect size is 0.6 and the intra-class correlation is 0.1. What is the power for them to find a significant difference between students from classrooms using the standard method and the new teaching method?

The input and output for calculating the power for this study are given in Figure 12.1.2. In the field of *Sample size*, input 20, the number of students within each class; in the field of *Effect size*, input 0.6, the expected effect size; in the field of *Number of clusters,* input 10, the number of classrooms; and in the field of *Intra-class correlation*, input 0.1, the expected intra-class correlation. The field for *Power* is left blank because it will be calculated. The default *Significance level* 0.05 is used although one can change it to a different value. The default *Type of analysis* is a two-sided test and one can change it to one-sided. By clicking the "**Calculate**" button, we get a power 0.5902 for the current study design.

Cluster randomized trials with 2 arms

Figure 12.1.2: Input and output for calculating power for CRT with 2 arms in Example 12.1.1

Parameters (Help)

Sample size   20

Effect size (Calculator) 0.6

Number of clusters   10

Intra-class correlation 0.1

Power

Significance level   0.05

Power curve   No power curve ↕

H1   Two-sided test ↕

Note   Cluster randomized tria

Calculate

Output

```
Cluster randomized trials with 2 arms

    J  n   f icc  power alpha
   10 20 0.6 0.1 0.5902  0.05

NOTE: n is the number of subjects per cluster.
URL: http://psychstat.org/crt2arm
```

A power curve is a line plot of statistical power along with given sample sizes. In Example 12.1.1, the power is 0.5902 with 20 students in 10 classrooms. What is the power for a different sample size, say, 40 in each class? And what is the power for a different cluster size, say 20 classrooms? One can investigate the power of different sample or cluster sizes and plot a power curve.

Example 12.1.2: Power curve with different sample sizes or cluster sizes

The input and output for plotting a power curve for the study in Example 12.1.1 are given in Figure 12.1.3. The sample size ranges from 20 to 80 with an interval of 10. In the *Sample size* field, the input is 20:80:10. We also choose "*Show power curve*" from the drop-down menu of *Power curve*. In the output, the power for each sample size from 20 to 80 with the interval 10 is listed. The power curve is displayed at the bottom of the output as shown in Figure 12.1.4. The power curve can be used for interpolation. For example, to get a power of 0.7, about 80 students are needed for each class. Similarly, one can obtain a power curve with different cluster sizes by specifying a sequence of numbers in the field *Number of clusters*.

Cluster randomized trials with 2 arms

Figure 12.1.3: Input and output for power curve for CRT with 2 arms in Example 12.1.2

Parameters (Help)

Sample size	20:80:10
Effect size (Calculator)	0.6
Number of clusters	10
Intra-class correlation	0.1
Power	
Significance level	0.05
Power curve	Show power curve ↕
H1	Two-sided test ↕
Note	Cluster randomized tria

Calculate

Output

```
Cluster randomized trials with 2 arms

   J   n   f icc  power alpha
  10  20 0.6 0.1 0.5902  0.05
  10  30 0.6 0.1 0.6365  0.05
  10  40 0.6 0.1 0.6620  0.05
  10  50 0.6 0.1 0.6781  0.05
  10  60 0.6 0.1 0.6891  0.05
  10  70 0.6 0.1 0.6971  0.05
  10  80 0.6 0.1 0.7032  0.05

NOTE: n is the number of subjects per cluster.
URL: http://psychstat.org/crt2arm
```

In practice, a power of 0.8 is often desired. Given the power, the number of clusters can be calculated as shown in Figure 12.1.5. In this situation, the *Number of clusters* field is left blank while the input for *Power* field is 0.8. In the output, we can see 15 classrooms are needed to obtain a power of 0.8. One can also obtain the required sample size by leaving *Sample size* field blank and specifying a cluster size in the *Number of clusters* field.

Example 12.1.3: Calculate sample size (or cluster size) given power, effect size and intra-class correlation

Figure 12.1.4: Power curve for CRT with 2 arms in Example 12.1.2

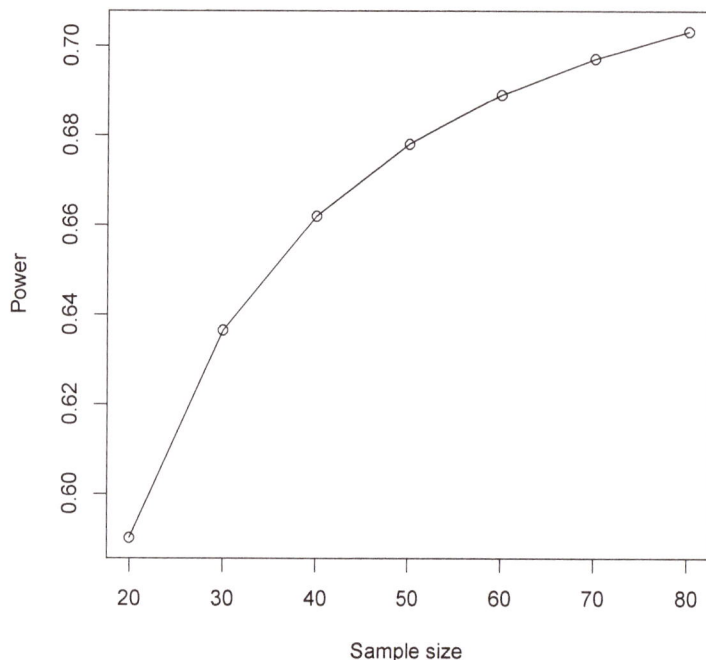

12.2 Use R Package WebPower

The power calculation for CRT with two arms is conducted using an R function `wp.crt2arm` function. The detail of the function is:

```
wp.crt2arm(n=NULL, f=NULL, J=NULL, icc=NULL, power=NULL,
    alternative = c("two.sided", "one.sided"), alpha=0.05)
```

n: sample size
f: effect size
J: number of clusters/sites
icc: Intra-class correlation
alpha: significance level
power: statistical power
alternative: two-sided or one-sided analysis

The R input and output for the examples in the above section are given below:

```
> ## calculate power given sample size and effect size
> wp.crt2arm(f=0.6,n=20,J=10,icc=.1)

Multilevel model cluster randomized trials with two arms

    J  n   f icc     power alpha
   10 20 0.6 0.1 0.5901684  0.05

NOTE: n is the number of observations in each cluster

WebPower URL: http://psychstat.org/crt2arm

>
> ## power curve
> res<-wp.crt2arm(f=0.6,n=seq(20,100,10),J=10,icc=.1)
```

Cluster randomized trials with 2 arms

Parameters (Help)

Sample size	20
Effect size (Calculator)	0.6
Number of clusters	
Intra-class correlation	0.1
Power	0.8
Significance level	0.05
Power curve	Show power curve ⬍
H1	Two-sided test ⬍
Note	Cluster randomized tria

Calculate

Output

```
Cluster randomized trials with 2 arms

       J  n    f icc power alpha
    14.84 20 0.6 0.1   0.8  0.05


NOTE: n is the number of subjects per cluster.
URL: http://psychstat.org/crt2arm
```

```
> res

Multilevel model cluster randomized trials with two arms

    J   n   f icc     power alpha
   10  20 0.6 0.1 0.5901684  0.05
   10  30 0.6 0.1 0.6365313  0.05
   10  40 0.6 0.1 0.6620030  0.05
   10  50 0.6 0.1 0.6780525  0.05
   10  60 0.6 0.1 0.6890755  0.05
   10  70 0.6 0.1 0.6971076  0.05
   10  80 0.6 0.1 0.7032181  0.05
   10  90 0.6 0.1 0.7080217  0.05
   10 100 0.6 0.1 0.7118967  0.05

NOTE: n is the number of observations in each cluster

WebPower URL: http://psychstat.org/crt2arm
```

```
> plot(res)  ## generate power curve
>
> ## number of clusters given effect size, sample size, power
> wp.crt2arm(f=0.6,n=20,J=NULL,icc=.1,power=.8)

Multilevel model cluster randomized trials with two arms

          J  n   f icc power alpha
   14.83587 20 0.6 0.1  0.8  0.05

NOTE: n is the number of observations in each cluster

WebPower URL: http://psychstat.org/crt2arm
```

12.3   Effect Size and Intra-class Correlation for CRT with 2 Arms

The effect size for CRT with two arms is defined as a ratio of the treatment main effect to the total standard deviation:

$$f = \frac{\mu_D}{\sqrt{\sigma_B^2 + \sigma_W^2}}, \tag{12.3.1}$$

where μ_D is the mean difference between the treatment clusters and the control clusters, σ_B^2 is the between-cluster variance and σ_W^2 is the within-cluster variance. Given the three quantities μ_D, σ_B^2 and σ_W^2, the effect size can be determined. The intra-class correlation (ICC) is defined as a ratio of the between-cluster variance to the total variance:

$$\rho = \frac{\sigma_B^2}{\sigma_B^2 + \sigma_W^2}. \tag{12.3.2}$$

The specification of the effect size and ICC can be assisted by an online calculator . In the interface in Figure 12.1.1, clicking the link "Calculator" brings up the calculator. The calculator, as shown in Figure 12.3.1, allows the following four methods of obtaining the effect size and ICC.

1. Input the mean difference μ_D, between cluster variance σ_B^2, and within cluster variance σ_W^2. This method calculates the effect size and ICC directly based on equations 12.3.1 and 12.3.2. With the three values, the effect size and ICC are calculated and shown after clicking on the "**Calculate**" button. Consider the example in 12.1.1 that previous research shows a new teaching method would lead to an increase of 2 in math scores. There are 10 classrooms with 20 students in each classroom. In addition, the between-classroom variance is 1 and the within-classroom variance is 4. To obtain the

effect size and ICC, the inputs for *Mean difference, Between cluster variance* and *Within cluster variance* are 2, 1, and 4, respectively. By clicking the "Calculate" button, the effect size and ICC are presented at the bottom as in Figure 12.3.1.

Effect Size Calculator for Two Level CRT with 2 Arms
1. Effect size parameter input

Mean difference	2
Between cluster variance	1
Within cluster variance	4
Sample size	
Number of clusters	
Means of clusters in treatment	
Means of clusters in control	
Ratio of within and between cluster variances	

Figure 12.3.1: Effect size and ICC calculation for CRT with 2 arms

Calculate

Effect size output

The effect size f = **0.8944**
The intra-class correlation ICC = **0.2000**

2. Input within cluster variance σ_W^2, sample size n, number of clusters J, and cluster means. Based on the literature, one might decide on the information for the cluster means as shown below.

Treatment		Control	
Cluster	mean	Cluster	mean
1	$\bar{Y}_{.1}^{T}$	1	$\bar{Y}_{.1}^{C}$
2	$\bar{Y}_{.2}^{T}$	2	$\bar{Y}_{.2}^{C}$
$\vdots$	$\bar{Y}_{.j}^{T}$	$\vdots$	$\bar{Y}_{.j}^{C}$
$J/2$	$\bar{Y}_{.\frac{J}{2}}^{T}$	$J/2$	$\bar{Y}_{.\frac{J}{2}}^{C}$

With such information, the mean difference can be calculated by

$$\hat{\mu_D} = \bar{Y}_{..}^{T} - \bar{Y}_{..}^{C}, \qquad (12.3.3)$$

where $\bar{Y}_{..}^{T} = \sum_{j=1}^{J/2} \bar{Y}_{.j}^{T}/(J/2)$, $\bar{Y}_{..}^{C} = \sum_{j=1}^{J/2} \bar{Y}_{.j}^{C}/(J/2)$. The between-cluster variance can be calculated by

$$\hat{\sigma_B}^2 = \frac{\sum_{j=1}^{J/2}(\bar{Y}_{.j}^{T} - \bar{Y}_{..}^{T})^2 + \sum_{j=1}^{J/2}(\bar{Y}_{.j}^{C} - \bar{Y}_{..}^{C})^2}{J-2} - \frac{\sigma_W^2}{n}. \qquad (12.3.4)$$

Consider the example that data are collected from 10 classrooms with 20 students in each classroom. The average scores for classrooms receiving treatment are 2, 4, 1, 3, 4; and those for classrooms in the control condition are 1, 2, 1, 3, 2. Further, the within-classroom variance is 4. The data can be input as in Figure 12.3.2. At the bottom, the effect size and ICC are presented.

Effect Size Calculator for Two Level CRT with 2 Arms
1. Effect size parameter input

Mean difference

Between cluster variance

Within cluster variance	4
Sample size	20
Number of clusters	10
Means of clusters in treatment	2 4 1 3 4
Means of clusters in control	1 2 1 3 2

Ratio of within and between cluster variances

Calculate

2. From empirical data analysis

Upload data file:
    Choose File No file chosen   Calculate

Effect size output

The effect size f = **0.4472**
The intra-class correlation ICC = **0.2000**

3. One can calculate ICC if he/she knows the ratio of within cluster variance to between cluster variance. With such information, the ICC can be calculated by

$$\rho = \frac{1}{1 + ratio}. \qquad (12.3.5)$$

Suppose one knows the ratio is 4. Then, in the field *Ratio of within and between cluster variances*, input 4. By clicking the "Calculate" button, the ICC is 0.2. Note that this approach should be used to calculate ICC only. The value for effect size is not valid under this approach.

4. The calculator also allows a user to upload a set of data and calculates effect size and ICC from the data directly.

Figure 12.3.2: Effect size and ICC calculation for CRT with 2 arms based on the input of cluster means

Example data:

ID	cluster	score	group
1	1	1	0
2	1	3	0
3	2	4	1
4	2	8	1
5	3	4	0
6	3	6	0
7	4	8	1
8	4	9	1

Figure 12.3.3 shows the use of the data in http://psychstat.org/crt2data and the output, including the estimated effect size of testing treatment main effect, estimated intra-class correlation, and the results from conducting a hypothesis test on the treatment main effect. Note that only registered users can use this method to protect data privacy. The data have to be in text format where the first column of the data is the ID variable, the second column represents the cluster, the third column is the outcome variable and the fourth column is the condition variable with 0 being the control condition and 1 being the treatment condition. The first line of the data should be the variable names.

Effect Size Calculator for Two Level CRT with 2 Arms
2. From empirical data analysis

Upload data file:

Choose File | CRT2.txt   Calculate

Effect size output

The effect size f = **1.0328**
The Intra-class correlation ICC = **0**

Test of fixed effects (X: Treatment main effect)

	Estimate	Std. Error	t value	p.value
(Intercept)	4.25	0.6038074	7.038669	0.01959328
X	2.00	1.2076147	1.656157	0.23953090

Figure 12.3.3: Effect size calculation based on an empirical set of data

12.4 How to Conduct Power Analysis for CRT with 3 Arms

The primary software interface for power analysis for CRT with three arms (i.e., two treatments and one control) is shown in Figure 12.4.1. Within the interface, a user can supply different parameter values and select different options for power analysis. Among the six parameters, *Sample size*, *Effect size*, *Number of clusters*, *Intra-class correlation*, *Significance level*, and *Power*, one and only one can be left blank.

http://psychstat.org/crt3arm

- The *Sample size* is the number of individuals within each cluster. The power calculation assumes a balanced design with equal sample size for each cluster. To obtain power estimation with varied sample sizes, multiple sample sizes can be provided in the following two ways.

First, multiple sample sizes can be supplied and separated by white spaces, e.g., 100 150 200 will calculate power for the three sample sizes 100, 150 and 200. Second, a sequence of sample (or cluster) sizes can be generated using the method `s:e:i` with `s` denoting the starting sample size, `e` as the ending sample size, and `i` as the interval. Note that the values are separated by colon ":". For example, 100:150:10 will generate a sequence of sample sizes: 100 110 120 130 140 150. By default, the sample size is 80.

- The *Effect size* quantifies the effect of a specific test. Three types of tests are available: 1) testing treatment main effect – the difference between the average treatment arms and the control arm; 2) testing the difference between the two treatment arms; and 3) testing whether the three arms are all equivalent. One can choose a test through *"Type of analysis"*. Multiple effect sizes or a sequence of effect sizes can be supplied using the same way for sample size. The default value is 0.6. One can either input the effect size directly or calculate it by clicking the "**Calculator**" in the parentheses. Note that the calculator also allows users to obtain intra-class correlation given the required information.

- The *Number of clusters* tells how many clusters are considered in the study design. More than three clusters are required. Multiple cluster sizes or a sequence of cluster sizes can be supplied using the same way for sample size.

- The *Intra-class correlation* is the ratio of between-cluster variance to the total variance, which quantifies the degree to which two randomly drawn observations within a cluster are correlated. The default value of intra-class correlation is 0.15. One can either input the intra-class correlation directly or calculate it by clicking the "**Calculator**" after *Effect size*.

- The *Power* specifies the desired statistical power, usually set at 0.80. The *H1* can be specified as "Two-sided test" and "One-sided test".

- The *Significance level* for power calculation is needed but usually is set at the default 0.05.

- In addition to the required input, one can also request the plot of a power curve if multiple sample sizes, or cluster sizes, or effect sizes are provided.

- A note (less than 200 characters) can also be provided to provide basic information on the analysis for future reference for registered users.

Once all fields have been appropriately filled in, clicking "**Calculate**" will create a table of output and, if specified, a power curve will appear below the table.

12.4.1 *Examples*

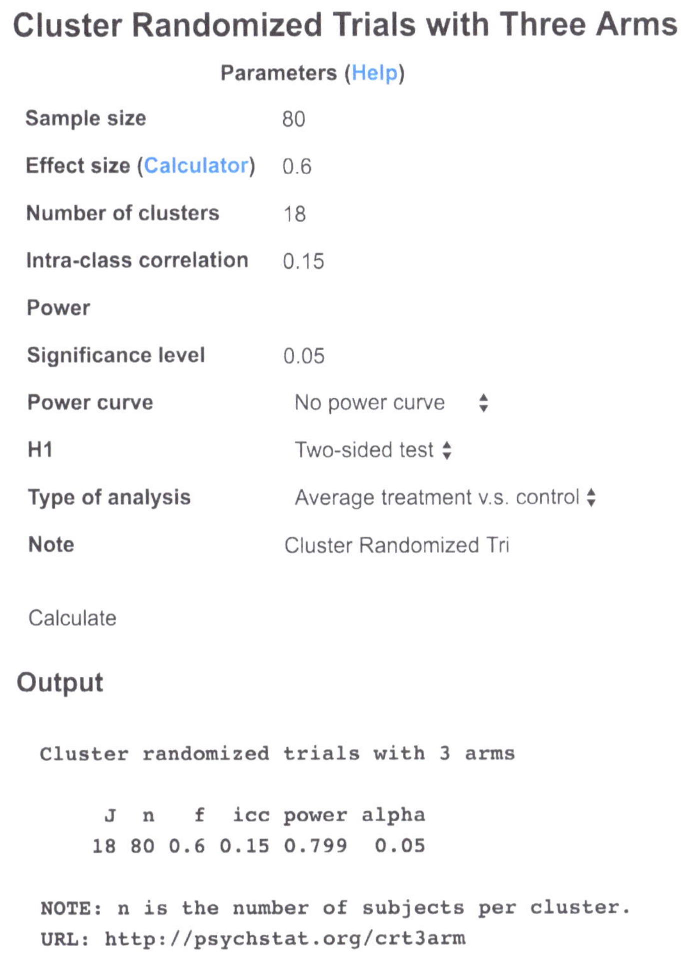

Cluster Randomized Trials with Three Arms

Figure 12.4.1: Software interface of power analysis for CRT with 3 arms

Parameters (Help)

Sample size	80
Effect size (Calculator)	0.6
Number of clusters	18
Intra-class correlation	0.15
Power	
Significance level	0.05
Power curve	No power curve ↕
H1	Two-sided test ↕
Type of analysis	Average treatment v.s. control ↕
Note	Cluster Randomized Tri

Calculate

Output

```
Cluster randomized trials with 3 arms

   J  n   f  icc power alpha
  18 80 0.6 0.15 0.799  0.05


NOTE: n is the number of subjects per cluster.
URL: http://psychstat.org/crt3arm
```

A medical researcher plans to compare two sleep aids and a placebo in helping people with sleep disorder. The outcome variable is self-reported sleep quality. The researcher plans to conduct the study in 21 clinics, with one-third receiving treatment 1, one-third receiving treatment 2, and the rest receiving placebo. Suppose there are 20 patients in each clinic. Past study reveals that the effect size for comparing the two sleep aids to the placebo is 0.5 and the intra-class correlation is 0.1. What is the power for detecting a significant difference between the average treatments and the placebo?

Example 12.4.1: Calculate power given sample sizes, effect size and intra-class correlation

The input and output for calculating power for this study are given in Figure 12.4.2. In the field of *Sample size*, input 20, the number of patients within each clinic; in the field of *Effect size*, input 0.5, the expected effect size; in the field of *Number of clusters*, input 21, the total number of clinics; and in the field of *Intra-class correlation*, input 0.1, the expected intra-class correlation. The field for *Power* is left blank because it will be calculated. The default *Significance level* 0.05 is used although one can change it to a different value. The default *H1* is a two-sided test and one can change it to one-sided if a one-sided test is to be conducted. In the field of *Type of analysis*, select "Average treatment v.s. control". By clicking the "**Calculate**" button, the statistical power 0.7651 is given in the output.

Cluster Randomized Trials with Three Arms

Figure 12.4.2: Input and output for calculating power for CRT with 3 arms in Example 12.4.1

Parameters (Help)

Sample size	20
Effect size (Calculator)	0.5
Number of clusters	21
Intra-class correlation	0.1
Power	
Significance level	0.05
Power curve	No power curve ⬍
H1	Two-sided test ⬍
Type of analysis	Average treatment v.s. control ⬍
Note	Cluster Randomized Tri

Calculate

Output

```
Cluster randomized trials with 3 arms

    J   n   f icc  power alpha
   21  20 0.5 0.1 0.7651  0.05


NOTE: n is the number of subjects per cluster.
URL: http://psychstat.org/crt3arm
```

A power curve is a line plot of statistical power along with given sample sizes. In Example 12.4.1, the power is 0.7651 with 20 patients in 21 clinics. What is the power for a different sample size, say, 40 in each

Example 12.4.2: Power curve with different sample sizes or cluster sizes

clinic? One can investigate the power of different sample or cluster sizes and plot a power curve.

The input and output for plotting a power curve for the study in Example 12.4.1 are given in Figure 12.4.3. The sample size ranges from 20 to 80 with an interval of 10. In the *Sample size* field, the input is 20:80:10. We also choose *"Show power curve"* from the drop-down menu of *Power curve*. In the output, the power for each sample size from 20 to 80 with the interval 10 is listed. The power curve is displayed at the bottom of the output as shown in Figure 12.4.4. The power curve can be used for interpolation. For example, to get a power of 0.8, about 30 patients are needed for each clinic. Similarly, one can obtain a power curve with different cluster sizes by specifying a sequence of numbers in the field *Number of clusters*.

In practice, a power of 0.8 is often desired. Given the power, the sample size can be calculated as shown in Figure 12.4.5. In this situation, the *Sample size* field is left blank while the input for *Power* field is 0.8. In the output, we can see 27 patients in each clinic are needed to obtain a power of 0.8. One can also obtain the required cluster size by leaving *Number of clusters* field blank and specifying a sample size in the *Sample size* field.

Example 12.4.3: Calculate sample size (or cluster size) given power, effect size and intra-class correlation

12.5 Use R Rackage WebPower

The power calculation for CRT with three arms is conducted using an R function wp.crt3arm function. The detail of the function is:

```
wp.crt3arm(n=NULL, f=NULL, J=NULL, icc=NULL, power=NULL, alpha
    =0.05, alternative = c("two.sided", "one.sided"), type=c("
    main","treatment","omnibus"))
```

The R input and output for the examples discussed in the previous section are given below:

```
> ## calculate power given sample size and effect size
> wp.crt3arm(n=20, f=.5, J=21, icc=.1, power=NULL)

Multilevel model cluster randomized trials with three arms

    J  n   f icc    power alpha
   21 20 0.5 0.1 0.7650611  0.05

NOTE: n is the number of observations in each cluster

WebPower URL: http://psychstat.org/crt3arm

>
> ## power curve
```

n: sample size
f: effect size
J: number of clusters/sites
icc: Intra-class correlation
alpha: significance level
power: statistical power
alternative: two-sided or one-sided analysis
type: main: the difference between the average treatment arms and the control arm, treatment: the difference between the two treatment arms, omnibus: the three arms are all equivalent.

```
> res<-wp.crt3arm(n=seq(20,100,10), f=.5, J=21, icc=.1, power=
    NULL)
> res

Multilevel model cluster randomized trials with three arms

    J   n   f icc      power alpha
   21  20 0.5 0.1 0.7650611  0.05
   21  30 0.5 0.1 0.8086098  0.05
   21  40 0.5 0.1 0.8309582  0.05
   21  50 0.5 0.1 0.8444483  0.05
   21  60 0.5 0.1 0.8534438  0.05
   21  70 0.5 0.1 0.8598585  0.05
   21  80 0.5 0.1 0.8646589  0.05
   21  90 0.5 0.1 0.8683838  0.05
   21 100 0.5 0.1 0.8713572  0.05

NOTE: n is the number of observations in each cluster

WebPower URL: http://psychstat.org/crt3arm

> plot(res)  ## generate power curve
>
> ## number of clusters given effect size, sample size, power
> wp.crt3arm(n=NULL, f=.5, J=21, icc=.1, power=.8)

Multilevel model cluster randomized trials with three arms

    J        n   f icc power alpha
   21 27.34175 0.5 0.1   0.8  0.05

NOTE: n is the number of observations in each cluster

WebPower URL: http://psychstat.org/crt3arm
```

12.6 *Effect Size and Intra-class Correlation for CRT with 3 Arms*

With a 3-arm CRT, one might be interested in three different types of test: (1) testing treatment main effect – the difference between the average treatment arms and the control arm, (2) testing the difference between the two treatment arms, and (3) testing whether the three arms are all equivalent (omnibus test). The corresponding effect sizes are

$$f_1 = \frac{\mu_{D1} + \mu_{D2}}{2\sqrt{\sigma_B^2 + \sigma_W^2}}, \tag{12.6.1}$$

$$f_2 = \frac{\mu_{D1} - \mu_{D2}}{\sqrt{\sigma_B^2 + \sigma_W^2}} \tag{12.6.2}$$

$$f_3 = \sqrt{\frac{\frac{1}{18}(\mu_{D1} + \mu_{D2})^2 + \frac{1}{6}(\mu_{D1} - \mu_{D2})^2}{\sigma_B^2 + \sigma_W^2}} \qquad (12.6.3)$$

where μ_{D1} is the mean difference between the first treatment and the control, μ_{D2} is the mean difference between the second treatment and the control, σ_B^2 is the between-cluster variance and σ_W^2 is the within-cluster variance. Given the four quantities μ_{D1}, μ_{D2}, σ_B^2 and σ_W^2, the effect sizes can be determined. The intra-class correlation (ICC) for all three tests is defined as a ratio of the between-cluster variance to the total variance:

$$\rho = \frac{\sigma_B^2}{\sigma_B^2 + \sigma_W^2}. \qquad (12.6.4)$$

The specification of the effect sizes and ICC can be assisted by an online calculator. In the interface in Figure 12.4.1, clicking the link "Calculator" brings up the calculator. The calculator, as shown in Figure 12.6.1, allows the following four methods of obtaining the effect size and ICC.

1. Input the mean differences μ_{D1} and μ_{D2}, between cluster variance σ_B^2, and within cluster variance σ_W^2. This method calculates the effect sizes and ICC directly based on equations 12.6.1 to 12.6.4. With the four values, the effect sizes and ICC are calculated and shown after clicking the "**Calculate**" button. Consider the example that it shows one sleep aid would increase 2 point scores in self-reported sleep quality, and the other sleep aid would increase 1 point score. The between-clinic variance is 1 and the within-clinic variance is 4. The data can be input as in Figure 12.6.1. At the bottom, the effect sizes for the three tests and the ICC are presented.

2. Input within-cluster variance σ_W^2, sample size n, number of clusters J, and cluster means. Based on the literature, one might decide on the information for the cluster means as shown below.

Treatment1		Treatment2		Control	
Cluster	mean	Cluster	mean	Cluster	mean
1	$\bar{Y}_{.1}^{T1}$	1	$\bar{Y}_{.1}^{T2}$	1	$\bar{Y}_{.1}^{C}$
2	$\bar{Y}_{.2}^{T1}$	2	$\bar{Y}_{.2}^{T2}$	2	$\bar{Y}_{.2}^{C}$
$\vdots$	$\bar{Y}_{.j}^{T1}$	$\vdots$	$\bar{Y}_{.j}^{T2}$	$\vdots$	$\bar{Y}_{.j}^{C}$
$J/3$	$\bar{Y}_{.\frac{J}{3}}^{T1}$	$J/3$	$\bar{Y}_{.\frac{J}{3}}^{T2}$	$J/3$	$\bar{Y}_{.\frac{J}{3}}^{C}$

With such information, the mean differences can be calculated by

$$\hat{\mu}_{D1} = \bar{Y}_{..}^{T1} - \bar{Y}_{..}^{C}, \; \hat{\mu}_{D2} = \bar{Y}_{..}^{T2} - \bar{Y}_{..}^{C}, \qquad (12.6.5)$$

where $\bar{Y}_{..}^{T1} = \sum_{j=1}^{J/3} \bar{Y}_{.j}^{T1}/(J/3)$, $\bar{Y}_{..}^{T2} = \sum_{j=1}^{J/3} \bar{Y}_{.j}^{T2}/(J/3)$, $\bar{Y}_{..}^{C} = \sum_{j=1}^{J/3} \bar{Y}_{.j}^{C}/(J/3)$. The between-cluster variance can be calculated by

$$\hat{\sigma}_B^2 = \frac{\sum_{j=1}^{J/3}(\bar{Y}_{.j}^{T1} - \bar{Y}_{..}^{T1})^2 + \sum_{j=1}^{J/3}(\bar{Y}_{.j}^{T2} - \bar{Y}_{..}^{T2})^2 + \sum_{j=1}^{J/3}(\bar{Y}_{.j}^{C} - \bar{Y}_{..}^{C})^2}{J - 3} - \frac{\sigma_W^2}{n}.$$

(12.6.6)

Consider the example that data are collected from 15 clinics with 5 patients in each clinic. The average scores in clinics receiving treatment 1 are 1, 3, 2, 4, 5; those in clinics receiving treatment 2 are 2, 3, 4, 1, 2; and those in clinics under the control condition are 1, 1, 2, 2, 1. Furthermore, the within-clinic variance is 1. The data can be input as in Figure 12.6.2. At the bottom, the effect sizes and ICC are presented.

3. One can calculate ICC if he/she knows the ratio of within-cluster variance to between-cluster variance. With such information, the ICC can be calculated by

$$\rho = \frac{1}{1 + ratio}.$$

(12.6.7)

Suppose one knows that the ratio is 4. Then, in the field *Ratio of within and between cluster variances*, input 4. By clicking the "Calculate" button, the ICC is 0.2. Note that this approach should be used to calculate ICC only. The values for effect sizes are not valid under this approach.

4. The calculator

Cluster Randomized Trials with Three Arms

Figure 12.4.3: Input and output for power curve for CRT with 3 arms in Example 12.4.2

Parameters (Help)

Sample size	20:80:10
Effect size (Calculator)	0.5
Number of clusters	21
Intra-class correlation	0.1
Power	
Significance level	0.05
Power curve	Show power curve ⬍
H1	Two-sided test ⬍
Type of analysis	Average treatment v.s. control ⬍
Note	Cluster Randomized Tri

Calculate

Output

```
Cluster randomized trials with 3 arms

    J   n    f icc  power alpha
   21 20 0.5 0.1 0.7651  0.05
   21 30 0.5 0.1 0.8086  0.05
   21 40 0.5 0.1 0.8310  0.05
   21 50 0.5 0.1 0.8444  0.05
   21 60 0.5 0.1 0.8534  0.05
   21 70 0.5 0.1 0.8599  0.05
   21 80 0.5 0.1 0.8647  0.05

NOTE: n is the number of subjects per cluster.
URL: http://psychstat.org/crt3arm
```

Figure 12.4.4: Power curve for CRT with
3 arms in Example 12.4.2

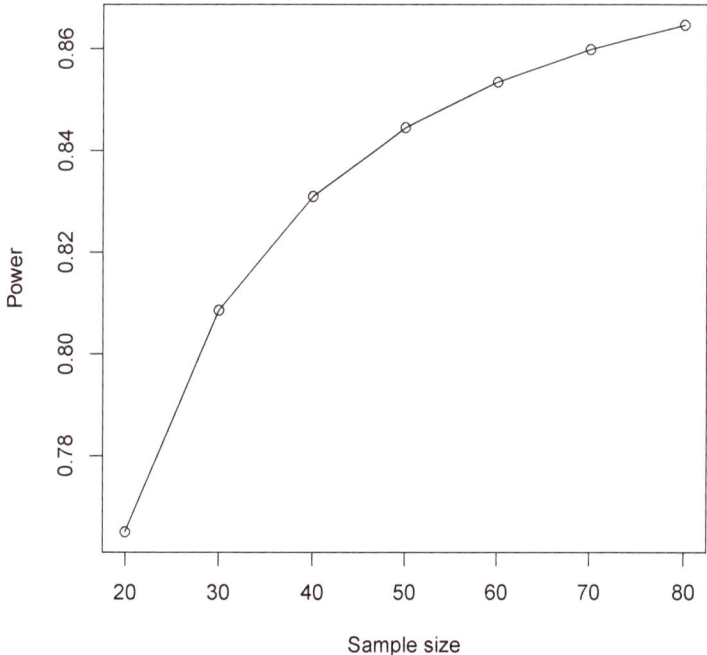

Cluster Randomized Trials with Three Arms

Figure 12.4.5: Input and output for sample size planning for CRT with 3 arms in Example 12.4.3

Parameters (Help)

Sample size

Effect size (Calculator)   0.5

Number of clusters   21

Intra-class correlation   0.1

Power   0.8

Significance level   0.05

Power curve   No power curve ⬍

H1   Two-sided test ⬍

Type of analysis   Average treatment v.s. control ⬍

Note   Cluster Randomized Tri

Calculate

Output

```
Cluster randomized trials with 3 arms

    J     n    f icc power alpha
   21 27.34 0.5 0.1   0.8  0.05

NOTE: n is the number of subjects per cluster.
URL: http://psychstat.org/crt3arm
```

Effect Size Calculator for Two Level CRT with 3 Arms
1. Effect size parameter input

Mean difference between treatment 1 and control   2

Mean difference between treatment 2 and control   1

Between cluster variance   1

Within cluster variance   4

Sample size

Number of clusters

Means of clusters in treatment 1

Means of clusters in treatment 2

Means of clusters in control

Ratio of within and between cluster variances

Calculate

2. From empirical data analysis

Upload data file:

Choose File   No file chosen   Calculate

Effect size output

The effect size for treatment main effect f1 = **0.6708**
The effect size for comparing the two treatments f2 = **0.4472**
The effect size for ominibus test f3 = **0.4830**
The intra-class correlation ICC = **0.2000**

Figure 12.6.1: Effect size and ICC calculation for CRT with 3 arms

Effect Size Calculator for Two Level CRT with 3 Arms
1. Effect size parameter input

Mean difference between treatment 1 and control	
Mean difference between treatment 2 and control	
Between cluster variance	
Within cluster variance	1
Sample size	5
Number of clusters	21
Means of clusters in treatment 1	1 3 2 4 5
Means of clusters in treatment 2	2 3 4 1 2
Means of clusters in control	1 1 2 2 1
Ratio of within and between cluster variances	

Calculate

2. From empirical data analysis

Upload data file:

Choose File | No file chosen   Calculate

Effect size output

The effect size for treatment main effect f1 = **0.9938**
The effect size for comparing the two treatments f2 = **0.4587**
The effect size for ominibus test f3 = **0.5045**
The intra-class correlation ICC = **0.4156**

Figure 12.6.2: Effect size and ICC calculation for CRT with 3 arms based on the input of cluster means

also allows a user to upload a set of data and calculate the sample effect size and ICC from data directly. Figure 12.6.3 shows the use of the data in http://psychstat.org/crt3data and the output of the effect sizes. Note that only registered users can use this method to protect data privacy. The data has to be in the plain text format where the first column of the data is the ID variable, the second column represents cluster, the third column is the outcome variable and the fourth column is the condition variable with 0 being the control condition, 1 being the first treatment and 2 being the second treatment. The first line of the data should be the variable names.

The output of Method 4 includes the estimated intra-class correlation and the estimated effect sizes of testing treatment main effect, comparing two treatments, and the omnibus test. In addition, this method also conducts hypothesis tests based on the data uploaded and presents the output as shown in Figure 12.6.3.

Example data:

```
ID  cluster  score  group
1        1    1.60      0
2        1    1.95      0
3        2    1.96      0
4        2    0.83      0
5        3    2.13      1
6        3    4.66      1
7        4    4.83      2
8        4    5.12      2
```

Effect Size Calculator for Two Level CRT with 3 Arms
2. From empirical data analysis

Upload data file:
  Choose File | CRT3.txt   Calculate

Figure 12.6.3: Effect size calculation based on an empirical set of data

Effect size output

The effect size of treatment main effect f1 = **0.6327**
The effect size of comparing two treatments f2 = **-0.6129**
The effect size of omnibus test f3 = **0.3893**
The Intra-class correlation ICC = **0**

Test of fixed effects (X: Treatment main effect; X1-X2: Comparing

```
            Estimate Std. Error  t value      p.value
(Intercept)   2.6480  0.1786612 14.82135  0.0006664012
X             0.6225  0.3789977  1.64249  0.1990301602
X1-X2        -0.6030  0.4376288 -1.37788  0.2620307107
Omnibust test: p-value = 0.09895558
```

12.7   *Technical Details*

In this section, we present the models for 2-arm CRT and 3-arm CRT, and formulas for power estimation under the two models. Details leading to power calculation can be found in Raudenbush (1997) and Liu (2013).

The model for a 2-arm CRT can be expressed as

$$Y_{ij} = \beta_{0j} + e_{ij}, \; e_{ij} \sim N(0, \sigma_W^2) \tag{12.7.1}$$

$$\beta_{0j} = \gamma_{00} + \gamma_{01} X_j + u_{0j}, \; u_{0j} \sim N(0, \sigma_B^2) \tag{12.7.2}$$

where Y_{ij} is the ith outcome in the jth cluster ($i = 1, 2, ..., N; \; j = 1, 2, ...J$), X_j is a treatment indicator of cluster j (0.5 for treatment and -0.5 for control), γ_{00} is the grand mean, γ_{01} is the treatment main effect that can be estimated by the difference between the averaged cluster mean in the treatment group and the averaged cluster mean in the control group, β_{0j} is the cluster mean for cluster j, σ_W^2 represents the within-cluster variance, and σ_B^2 represents the between-cluster variance.

The test for treatment main effect uses a t-test. Under $H_0 : \gamma_{01} = 0$, the test statistic follows a central t distribution with the degrees of freedom $J - 2$. Under the alternative hypothesis, the test statistic follows a non-central t distribution with the degrees of freedom $J - 2$ and the non-central parameter λ, which takes the form

$$\lambda = \frac{\gamma_{01}}{\sqrt{4(\sigma_B^2 + \frac{\sigma_W^2}{\sigma_B^2})/J}}. \tag{12.7.3}$$

Therefore, the power for the t-test of the treatment effect can be calculated by

$$\text{Power} = \begin{cases} 1 - P[T_{J-2,\lambda} < t_0] + P[T_{J-2,\lambda} \leq -t_0] & \text{two} - \text{sided test,} \\ 1 - P[T_{J-2,\lambda} < t_0] & \text{one} - \text{sided test,} \end{cases} \tag{12.7.4}$$

where t_0 is the $100(1 - \frac{\alpha}{2})$th percentile of the t distribution with $J - 2$ degrees of freedom for a two-sided test, and is the $100(1 - \alpha)$th percentile of the t distribution with $J - 2$ degrees of freedom for a one-sided test; and α is the significance level.

The model for a 3-arm CRT can be expressed as

$$Y_{ij} = \beta_{0j} + e_{ij}, \; e_{ij} \sim N(0, \sigma_W^2) \tag{12.7.5}$$

$$\beta_{0j} = \gamma_{00} + \gamma_{01} X_{1j} + \gamma_{02} X_{2j} + u_{0j}, \; u_{0j} \sim N(0, \sigma_B^2) \tag{12.7.6}$$

where Y_{ij} is the ith outcome in the jth cluster ($i = 1, 2, ..., N; \; j = 1, 2, ...J$), X_{1j} is used to compare the average outcome of the two treatment arms with that of the control arm (1/3 for the first treatment, 1/3 for the second treatment and -2/3 for the control condition), X_{2j} is used to contrast the average outcome between the two treatment arms (1/2 for the first treatment, -1/2 for the second treatment and 0 for the control condition), β_{0j} is the cluster mean, γ_{00} is the grand mean, γ_{01} is the treatment main effect that can be estimated by the difference between the average of the two treatment conditions and the control condition, and γ_{02} is the mean difference between the two treatment arms. Power for testing treatment main effect $H_0 : \gamma_{01} = 0$ is calculated by

$$
\text{Power} = \begin{cases} 1 - P[T_{J-3,\lambda_1} < t_0] + P[T_{J-3,\lambda_1} \le -t_0] & two-sided\ test, \\ 1 - P[T_{J-3,\lambda_1} < t_0] & one-sided\ test, \end{cases}
$$

$$(12.7.7)$$

where

$$
\lambda_1 = \frac{\gamma_{01}}{\sqrt{4.5(\sigma_B^2 + \frac{\sigma_W^2)}{n})/J}}, \tag{12.7.8}
$$

t_0 is the $100(1 - \frac{\alpha}{2})$th percentile for a two-sided test, and is the $100(1 - \alpha)$th percentile for a one-sided test of the t distribution with $J - 3$ degrees of freedom. Power for testing the difference between two treatments $H_0 : \gamma_{02} = 0$ is calculated by

$$
\text{Power} = \begin{cases} 1 - P[T_{J-3,\lambda_2} < t_0] + P[T_{J-3,\lambda_2} \le -t_0] & two-sided\ test, \\ 1 - P[T_{J-3,\lambda_2} < t_0] & one-sided\ test, \end{cases}
$$

where

$$
\lambda_2 = \frac{\gamma_{02}}{\sqrt{6(\sigma_B^2 + \frac{\sigma_W^2)}{n})/J}}, \tag{12.7.9}
$$

t_0 is defined as in Equation 12.7.7. Power for the omnibus test $H_0 : \gamma_{01} = \gamma_{02} = 0$ is calculated by

$$
\text{Power} = P(F_{2,J-3,\lambda_3} \ge F_0), \tag{12.7.10}
$$

based on the central and non-central F distribution where

$$
\lambda_3 = \lambda_1^2 + \lambda_2^2 = \frac{\gamma_{01}^2}{4.5(\sigma_B^2 + \frac{\sigma_W^2)}{n})/J} + \frac{\gamma_{02}^2}{6(\sigma_B^2 + \frac{\sigma_W^2)}{n})/J}, \tag{12.7.11}
$$

F_0 is the $100(1 - \alpha)$th percentile of the F distribution with degrees of freedom 2 and $J - 3$.

12.8   Exercises

1. A medical investigator is interested in whether a newly developed drug could lower blood pressure for patients with hypertension.

He plans to collect data from 4 clinics, with patients in 2 clinics to receive the new drug and those in the other 2 clinics to be in the control condition. Each clinic is expected to have 20 patients. The investigator hypothesizes that the new drug could lower patient's blood pressure by 8mmHg. The between-cluster variance is 2 and the within-cluster variance is 4. What is the effect size of treatment effect? What is the power for the investigator to find a significant difference in patients who use the new drug and those who do not?

2. Using the same information in Exercise 1, what would be the required sample sizes when the alpha level is set at 0.1? at 0.01?

3. Using the same information in Exercise 1, generate a power curve with the cluster size ranging from 4 to 8 with an interval of 1. From the power curve, approximately how large a cluster size is needed to get a power 0.9?

4. Using the same information in Exercise 1, generate a power curve with the sample size ranging from 20 to 200 with an interval of 20.

13   *Statistical Power Analysis for Multisite Randomized Trials*

Miao Yang
Department of Psychology
University of Notre Dame

Multilevel designs for cross-sectional studies typically contain cluster randomized trials and multisite randomized trials (Liu, 2013). Multisite randomized trials (MRT) are used when individuals within a cluster are randomly assigned to either a treatment arm or a control arm. The data from MRT can be analyzed in a two-level hierarchical linear model, where the indicator variable for treatment assignment is included in the first level. If a study contains multiple treatments, then multiple indicators will be used. Our power analysis for MRT focuses on designs with 2 arms (i.e., a treatment and a control) and 3 arms (i.e., two treatments and a control).

13.1   *How to Conduct Power Analysis for MRT with 2 Arms*

The primary software interface for power analysis for MRT with two arms (i.e., one treatment and one control at each site) is shown in Figure 13.1.1. Within the interface, a user can supply different parameter values and select different options for power analysis. In a 2-arm MRT, one might be interested in the following three types of tests:

http://psychstat.org/mrt2arm

- Testing treatment main effect.

  - To obtain power for this test, users need to input *Sample size, Effect size, Number of clusters, Variance of treatment effects across sites, Level-one error variance, Significance level.*

- Testing site variability.

  - To obtain power for this test, users need to input *Sample size,*

Number of clusters, Variance of site means, Level-one error variance, Significance level.

• Testing the variance of treatment main effect.

  – To obtain power for this test, users need to input *Sample size, Number of clusters, Variance of treatment effects across sites, Level-one error variance, Significance level.*

Multisite randomized trials with 2 arms

Figure 13.1.1: Software interface of power analysis for MRT with 2 arms

Parameters (Help)

Sample size	100
Effect size (Calculator)	0.6
Number of clusters	25
Variance of site means	1
Variance of treatment effects	1
Level 1 error variance	1
Power	
Significance level	0.05
Power curve	No power curve ⬍
H1	Two-sided test ⬍
Type of test	Main effect ⬍
Note	Multisite randomized tri:

Calculate

More information about each input is given below.

• The *Sample size* is the number of individuals within each cluster. The power calculation only allows for balanced data. In other words, all clusters contain an equal number of subjects and within each cluster, an equal number of subjects are assigned to the treatment condition and the control condition. To obtain power estimation with varied sample sizes, multiple sample sizes can be provided in the following two ways. First, multiple sample sizes can be supplied and separated by white spaces, e.g., 100 150 200 will calculate power for the three sample sizes 100, 150 and 200. Second, a sequence of sample (or cluster) sizes can be generated using the method `s:e:i` with s denoting the starting sample size, e as the ending sample size, and i as the interval. Note that the values are separated by colon

":". For example, 100:150:10 will generate a sequence of sample sizes: 100 110 120 130 140 150. By default, the sample size is 80.

- The *Effect size* specifies the main effect of treatment. Multiple effect sizes or a sequence of effect sizes can be supplied using the same way for the sample size. The default value is 0.6. One can either input the effect size directly or calculate it by clicking the "**Calculator**" in the parentheses. Note that only the first test (testing treatment main effect) requires an input for effect size.

- The *Number of clusters* tells how many clusters are considered in the study design. At least two clusters are required.

- The *Variance of site means* and the *Variance of treatment effects across sites* are the level-2 residual variances.

- The *Level-one error variance* is the level-1 residual variance.

- The *Power* specifies the desired statistical power, usually set at 0.8.

- The *Significance level* for power calculation is needed but is usually set at the default 0.05. The *H1* can be specified as "Two-sided test" and "One-sided test".

- The *Type of test* specifies which test is conducted.

- In addition to the required input, one can also request the plot of a power curve if multiple sample sizes, or cluster sizes, or effect sizes are provided.

- A note (less than 200 characters) can also be provided to provide basic information on the analysis for future reference for registered users.

Once all fields have been appropriately filled in, clicking "**Calculate**" will create a table of output and, if specified, a power curve will appear below the table.

13.1.1   *Examples*

A researcher plans to conduct a multisite randomized trial to evaluate the efficacy of an intervention for alcohol abuse. Patients will be recruited from 20 sites, and at each site half of the patients will be assigned to the treatment condition and the other half will be assigned to the control condition. The number of patients at each site is expected to be 40. The outcome variable is the reduction in abuse symptoms. A past study reveals that the effect size for the treatment is 0.5. Further, the researcher estimates that the variance for individual error is 1.25,

Example 13.1.1: Calculate power given sample size and effect size

the variance of site means is 0.1 and the variance in treatment effect across sites is 0.5. What's the power for testing the treatment main effect? What's the power for testing the variance of treatment main effect across sites? What's the power for testing the site variability?

The input and output for calculating power for testing the treatment main effect are given in Figure 13.1.2. In the field of *Sample size*, input 45, the number of patients within each site; in the field of *Effect size*, input 0.5, the expected effect size; in the field of *Number of clusters*, input 20, the number of sites; in the field of *Variance of treatment effects across sites*, input 0.5; and in the field of *Level-one error variance*, input 1.25. The field for *Power* is left blank because it will be calculated. The default *Significance level* 0.05 is used although one can change it to a different value. The default *H1* is a two-sided test. In the field of *Type of analysis*, select "Treatment main effect". By clicking the "**Calculate**" button, the statistical power is given in the output, and for the current design, the power is 0.8506.

The input and output for calculating power for testing the variance of treatment main effect across sites are given in Figure 13.1.3. In the field of *Sample size*, input 40; in the field of *Number of clusters*, input 20; in the field of *Variance of treatment effects across sites*, input 0.5; and in the field of *Level-one error variance*, input 1.25. The field for *Power* is left blank because it will be calculated. The default *Significance level* 0.05 is used. The default *H1* is a two-sided test. In the field of *Type of analysis*, select "Variance of treatment effects across sites". By clicking the "**Calculate**" button, the statistical power is given in the output: 0.9976.

The input and output for calculating power for testing site variability are given in Figure 13.1.4. In the field of *Sample size*, input 45; in the field of *Number of clusters*, input 20; in the field of *Variance of site means*, input 0.1; and in the field of *Level-one error variance*, input 1.25. The field for *Power* is left blank because it will be calculated. The default *Significance level* 0.05 is used. The default *H1* is a two-sided test. In the field of *Type of analysis*, select "Site variability". By clicking the "**Calculate**" button, the statistical power is given in the output as 0.9925.

Example 13.1.2: Power curve with different sample sizes or cluster sizes

The input and output for plotting a power curve for testing the treatment main effect in Example 13.1.1 are given in Figure 13.1.5. The sample size of each site ranges from 10 to 50 with an interval of 10. In the *Sample size* field, the input is 10:50:10. We also choose "*Show power curve*" from the drop-down menu of *Power curve*. In the output, the power for each sample size from 10 to 50 with the interval 10 is listed. The power curve is displayed at the bottom of the output as shown in Figure 13.1.6. The power curve can be used for interpolation. For example, to get a power 0.8, about 25 patients are needed for each site.

Multisite randomized trials with 2 arms

Figure 13.1.2: Input and output for calculating power for testing the treatment main effect using MRT with 2 arms in Example 13.1.1

Parameters (Help)

Sample size	40
Effect size (Calculator)	0.5
Number of clusters	20
Variance of site means	
Variance of treatment effects	0.5
Level 1 error variance	1.25
Power	
Significance level	0.05
Power curve	No power curve ⇕
H1	Two-sided test ⇕
Type of test	Main effect ⇕
Note	Multisite randomized tri

Calculate

Output

```
Multisite randomized trials with 2 arms

    J   n   f tau11  sg2  power alpha
   20  40 0.5    0.5 1.25 0.8506  0.05


NOTE: n is the number of subjects per cluster
URL: http://psychstat.org/mrt2arm
```

Similarly, one can obtain a power curve with different cluster sizes by specifying a sequence of numbers in the field *Number of clusters*.

In practice, a power of 0.8 is often desired. Given the power, the sample size can be calculated as shown in Figure 13.1.7. In this situation, the *Sample size* field is left blank while the input for the *Power* field is 0.8. In the output, we can see 24 patients are needed for each site to obtain a power of 0.8 for testing treatment main effect. One can also obtain the required site size by leaving *Number of clusters* field blank and specifying a sample size in the *Sample size* field.

Example 13.1.3: Calculate sample size (or cluster size) given power and effect size

Multisite randomized trials with 2 arms

Parameters (Help)

Sample size	40
Effect size (Calculator)	
Number of clusters	20
Variance of site means	
Variance of treatment effects	0.5
Level 1 error variance	1.25
Power	
Significance level	0.05
Power curve	No power curve $\updownarrow$
H1	Two-sided test $\updownarrow$
Type of test	Variance of treatment effects $\updownarrow$
Note	Multisite randomized tri:

Calculate

Figure 13.1.3: Input and output for calculating power for testing variance of treatment effects across sites using MRT with 2 arms in Example 13.1.1

Output

```
Multisite randomized trials with 2 arms

   J  n tau11  sg2  power alpha
  20 40   0.5 1.25 0.9976  0.05


NOTE: n is the number of subjects per cluster
URL: http://psychstat.org/mrt2arm
```

13.2   Use R Package WebPower

The power calculation for MRT with two arms is conducted using an R function `wp.mrt2arm` function. The detail of the function is:

```
wp.mrt2arm(n=NULL, f=NULL, J=NULL, tau00=NULL, tau11=NULL, sg2=
    NULL, power=NULL, alpha=0.05, alternative = c("two.sided", "
    one.sided"), type=c("main","site","variance"))
```

The R input and output for the examples used in the previous section are given below:

```
> ## calculate power for main effect given sample size and effect
    size
> wp.mrt2arm(n=45, f=0.5, J=20, tau11=.5, sg2=1.25, power=NULL)

Multilevel model multisite randomized trials with two arms
```

n: sample size
f: effect size
J: number of clusters/sites
tau00: Variance of site means
tau11: Variance of treatment effects across sites
sg2: Level-one error variance
alpha: significance level
power: statistical power
alternative: two-sided or one-sided analysis
type: main: testing treatment main effect; site: testing site variability; variance: testing the variance of treatment main effect.

Multisite randomized trials with 2 arms

Figure 13.1.4: Input and output for calculating power for testing site variability using MRT with 2 arms in Example 13.1.1

Parameters (Help)

Sample size 40

Effect size (Calculator)

Number of clusters 20

Variance of site means 0.1

Variance of treatment effects

Level 1 error variance 1.25

Power

Significance level 0.05

Power curve No power curve ↕

H1 Two-sided test ↕

Type of test Site variability ↕

Note Multisite randomized tri

Calculate

Output

```
Multisite randomized trials with 2 arms

    J  n tau00  sg2  power alpha
   20 40   0.1 1.25 0.9925  0.05


NOTE: n is the number of subjects per cluster
URL: http://psychstat.org/mrt2arm
```

```
    J  n   f tau11  sg2      power alpha
   20 45 0.5   0.5 1.25 0.8583253  0.05


NOTE: n is the number of observations in each cluster

WebPower URL: http://psychstat.org/mrt2arm

> ## for testing variance of treatment effect
> wp.mrt2arm(n=45, J=20, tau11=.5, sg2=1.25, power=NULL, type="
    variance")

Multilevel model multisite randomized trials with two arms

    J  n tau11  sg2      power alpha
```

Multisite randomized trials with 2 arms

Parameters (Help)

Sample size	10:50:10
Effect size (Calculator)	0.5
Number of clusters	20
Variance of site means	
Variance of treatment effects	0.5
Level 1 error variance	1.25
Power	
Significance level	0.05
Power curve	Show power curve ⬍
H1	Two-sided test ⬍
Type of test	Main effect ⬍
Note	Multisite randomized tri:

Calculate

Figure 13.1.5: Input and output for power curve for testing the treatment main effect using MRT with 2 arms in Example 13.1.2

Output

```
Multisite randomized trials with 2 arms

    J   n    f tau11  sg2  power alpha
   20  10  0.5    0.5 1.25 0.6599  0.05
   20  20  0.5    0.5 1.25 0.7818  0.05
   20  30  0.5    0.5 1.25 0.8274  0.05
   20  40  0.5    0.5 1.25 0.8506  0.05
   20  50  0.5    0.5 1.25 0.8645  0.05

NOTE: n is the number of subjects per cluster
URL: http://psychstat.org/mrt2arm
```

```
   20 45    0.5 1.25 0.9987823  0.05

NOTE: n is the number of observations in each cluster

WebPower URL: http://psychstat.org/mrt2arm

> ## for testing site variability
> wp.mrt2arm(n=45, J=20, tau00=.1, sg2=1.25, power=NULL, type="
    site")

Multilevel model multisite randomized trials with two arms

    J  n tau00  sg2     power alpha
   20 45   0.1 1.25 0.9958889  0.05
```

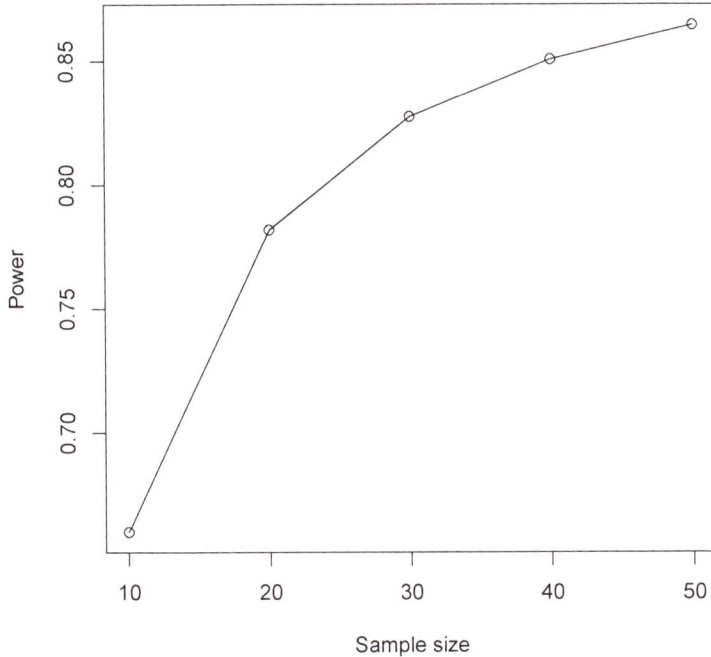

Figure 13.1.6: Power curve for testing the treatment main effect using MRT with 2 arms in Example 13.1.2

```
NOTE: n is the number of observations in each cluster

WebPower URL: http://psychstat.org/mrt2arm

>
> ## power curve
> res<-wp.mrt2arm(n=seq(10,50,5), f=0.5, J=20, tau11=.5, sg2
    =1.25, power=NULL)
> res

Multilevel model multisite randomized trials with two arms

   J  n   f tau11  sg2     power alpha
  20 10 0.5   0.5 1.25 0.6599499  0.05
  20 15 0.5   0.5 1.25 0.7383281  0.05
  20 20 0.5   0.5 1.25 0.7818294  0.05
  20 25 0.5   0.5 1.25 0.8090084  0.05
  20 30 0.5   0.5 1.25 0.8274288  0.05
  20 35 0.5   0.5 1.25 0.8406659  0.05
  20 40 0.5   0.5 1.25 0.8506049  0.05
  20 45 0.5   0.5 1.25 0.8583253  0.05
  20 50 0.5   0.5 1.25 0.8644864  0.05

NOTE: n is the number of observations in each cluster
```

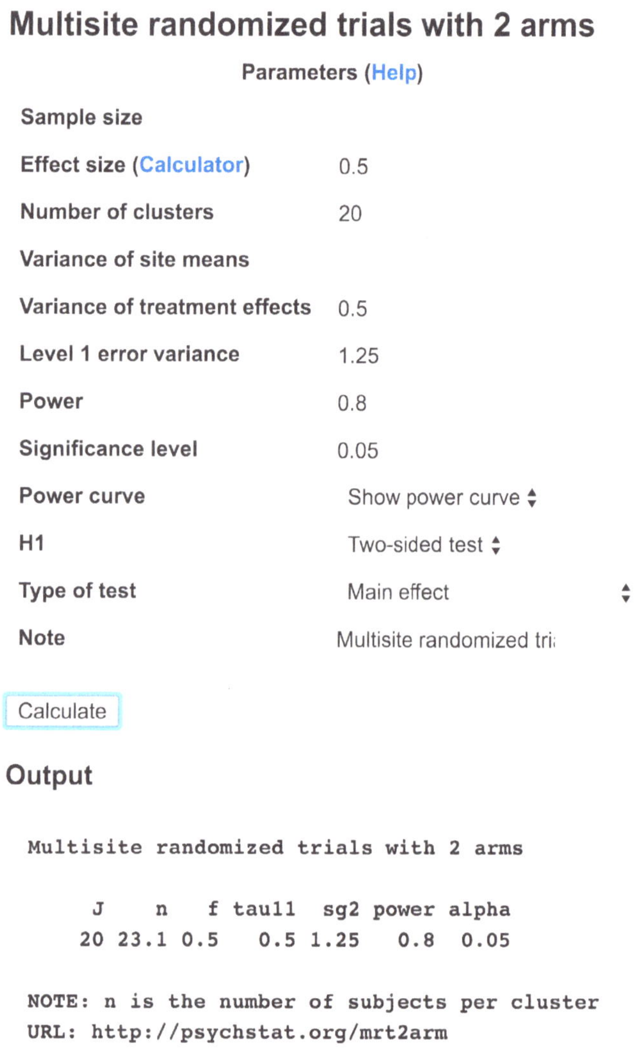

Multisite randomized trials with 2 arms

Figure 13.1.7: Input and output for sample size planning for MRT with 2 arms in Example 13.1.3

Parameters (Help)

Sample size

Effect size (Calculator)   0.5

Number of clusters   20

Variance of site means

Variance of treatment effects   0.5

Level 1 error variance   1.25

Power   0.8

Significance level   0.05

Power curve   Show power curve ▼

H1   Two-sided test ▼

Type of test   Main effect ▼

Note   Multisite randomized tri

Calculate

Output

```
Multisite randomized trials with 2 arms

    J    n    f tau11  sg2 power alpha
   20 23.1 0.5   0.5 1.25   0.8  0.05

NOTE: n is the number of subjects per cluster
URL: http://psychstat.org/mrt2arm




WebPower URL: http://psychstat.org/mrt2arm

> plot(res)  ## generate power curve
>
> ## sample size given effect size, sample size, power
> wp.mrt2arm(n=NULL, f=0.5, J=20, tau11=.5, sg2=1.25, power=.8)

Multilevel model multisite randomized trials with two arms

     J       n  f tau11  sg2 power alpha
    20 23.10086 0.5   0.5 1.25   0.8  0.05

NOTE: n is the number of observations in each cluster
```

WebPower URL: http://psychstat.org/mrt2arm

13.3   *Effect Size for MRT with 2 Arms*

As noted previously, there are three types of tests that are of interest in a 2-arm MRT: testing treatment main effect, testing site variability, and testing the variance of treatment main effect. The corresponding effect sizes are defined as

$$f_1 = \frac{\mu_D}{\sqrt{\sigma^2}}, \qquad (13.3.1)$$

$$f_2 = \frac{\tau_{00}}{\sigma^2}, \qquad (13.3.2)$$

$$f_3 = \frac{\tau_{11}}{\sigma^2}, \qquad (13.3.3)$$

where μ_D is the mean difference between the treatment and control across all the sites, σ^2 is the level-1 error variance, τ_{00} is the variance of site means, and τ_{11} is the variance of treatment effects across sites.

Note that the *Effect size* in the interface in Figure 13.1.1 refers to the effect size for testing treatment main effect. Its specification can be assisted by an online calculator, as shown in Figure 13.3.1. A user needs to supply the *Mean difference* between the treatment and control as well as the *Level-one error variance*. Once these fields have been filled out, the user can click "Calculate" and the effect size will be given at the bottom.

Effect Size Calculator for Two Level MRT with 2 Arms
1. Effect size parameter input

Mean difference

Level 1 error variance

Calculate

2. From empirical data analysis

Upload data file:

Choose File   No file chosen   Calculate

Alternatively,
one can upload a set of empirical data to obtain the effect size. Only registered users can use this method to protect data privacy. The data have to be in the plain text format where the first column of the data is the ID variable, the second column represents a cluster

Figure 13.3.1: Effect size calculator for MRT with 2 arms

Example data:

ID	cluster	score	group
1	1	1	0
2	1	2	0
3	1	4	1
4	1	6	1
5	2	3	0
6	2	3	0
7	2	2	1
8	2	6	1

variable, the third column is the outcome variable and the fourth column is the condition variable with 0 being the control condition and 1 being the treatment condition. The first line of the data should be the variable names. Figure 13.3.2 shows the use of the data online at http://psychstat.org/mrt2data and the output – including the estimated effect size of testing treatment main effect and the results from conducting the hypothesis testing.

Effect Size Calculator for Two Level MRT with 2 Arms

2. From empirical data analysis

Upload data file:

Choose File | MRT2.txt   Calculate

Effect size output

The effect size of treatment main effect f = **-0.2987**

Test of fixed effects (X: Treatment main effect)

```
            Estimate Std. Error    t value     p.value
(Intercept)   4.9375  0.8156494  6.0534586  0.009043958
X            -1.1250  1.6093474 -0.6990411  0.534848038
```

Figure 13.3.2: Effect size calculation based on an empirical set of data

13.4   *How to Conduct Power Analysis for MRT with 3 Arms*

The primary software interface for power analysis for MRT with three arms (i.e., two treatments and one control at each site) is shown in Figure 13.4.1. Within the interface, a user can supply different parameter values and select different options for power analysis. In a 3-arm MRT, one might be interested in the following three types of tests.

http://psychstat.org/mrt3arm

- Testing treatment main effect: whether there is a difference between the average treatment arms and the control arm.

  – To obtain power for this test, users need to input *Sample size, Effect size of treatment main effect, Number of clusters, Variance of treatment effects across sites, Level-one error variance, Significance level,* and select "Test treatment main effect" after *Type of analysis.*

- Testing the difference between the two treatments.

  – To obtain power for this test, users need to input *Sample size, Effect size of two treatment differences, Number of clusters, Variance of*

treatment effects across sites, Level-one error variance, Significance level,
and select "Compare two treatments" after *Type of analysis.*

- Omnibus test: testing whether there is any difference among all
  three arms.

  - To obtain power for this test, users need to input *Sample size, Effect
    size of treatment main effect, Effect size of two treatment differences,
    Number of clusters, Variance of treatment effects across sites, Level-one
    error variance, Significance level,* and select "Omnibus test" after
    Type of analysis.

Multisite randomized trials with 3 arms

Figure 13.4.1: Software interface of power analysis for MRT with 3 arms

Parameters (Help)	
Sample size	100
Effect size of treatment main effect (Calculator)	0.8
Effect size of two treatment difference	0.8
Number of clusters	20
Variance of treatment effects	1
Level 1 error variance	1
Power	
Significance level	0.05
Power curve	No power curve ⬍
H1	Two-sided test ⬍
Type of analysis	Main effect ⬍
Note	Multisite randomized tri:

Calculate

The details about the input are given below.

- The *Sample size* is the number of individuals within each cluster. The
  power calculation only allows for balanced data. In other words, all
  clusters contain an equal number of subjects and within each cluster,
  an equal number of subjects are assigned to the three conditions. To
  obtain power estimation with varied sample sizes, multiple sample
  sizes can be provided in the following two ways. First, multiple
  sample sizes can be supplied and separated by white spaces, e.g.,
  100 150 200 will calculate power for the three sample sizes 100,

150 and 200. Second, a sequence of sample (or cluster) sizes can be generated using the method s:e:i with s denoting the starting sample size, e as the ending sample size, and i as the interval. Note that the values are separated by colon ":". For example, 100:150:10 will generate a sequence of sample sizes: 100 110 120 130 140 150. By default, the sample size is 80.

- The *Effect size of treatment main effect* and the *Effect size of two treatment difference* quantify the effect for comparing the average treatment arms to the control arm, and the effect for comparing the two treatment arms, respectively. Multiple effect sizes or a sequence of effect sizes can be supplied using the same way for sample size. One can either input the effect sizes directly or calculate them by clicking the "**Calculator**" in the parentheses.

- The *Number of clusters* tells how many clusters are considered in the study design. At least two clusters are required.

- The *Variance of treatment effects across sites* specifies the site variability for the treatments.

- The *Level-one error variance* is the level-1 residual variance.

- The *Power* specifies the desired statistical power, usually set at 0.80.

- The *H1* can be specified as "Two-sided test" and "One-sided test".

- The *Significance level* for power calculation is needed but usually set at the default 0.05.

- The *Type of test* specifies which test is to be conducted.

- In addition to the required input, one can also request the plot of a power curve if multiple sample sizes, or cluster sizes, or effect sizes are provided.

- A note (less than 200 characters) can also be provided to provide basic information on the analysis for future reference for registered users.

Once all fields have been appropriately filled in, pressing "Calculate" will create a table of output and, if specified, a power curve will appear below the table.

13.4.1 *Examples*

A researcher plans to collect data from 20 clinics to examine the effect of certain behavioral therapies on recovering from anorexia. At

Example 13.4.1: Calculate power given sample size and effect size

each clinic, 30 girls will be randomly assigned to therapy 1, therapy 2, or the control group. Previous research suggests that therapy 1 might lead to an increase of 0.5 in BMI and therapy 2 might lead to an increase of 0.8 in BMI. Further, the person-specific error variance is 2.25 and the variance in treatment effects across sites is 0.4. What's the power for testing the treatment main effect? What's the power for testing the difference between the two treatments? What's the power for the omnibus test?

Although the effect sizes are not provided, one can calculate them based on the given information. In the next subsection, we will define the effect sizes and illustrate how to obtain them by using the "**Calculator**". It turns out that the effect size of treatment main effect is 0.43 and that of two treatment difference is 0.2.

The input and output for calculating power for testing the treatment main effect are given in Figure 13.4.2. In the field of *Sample size*, input 30, the number of patients within each clinic; in the field of *Effect size of treatment main effect*, input 0.43, the expected effect size of treatment main effect; in the field of *Number of clusters*, input 20, the number of clinics; in the field of *Variance of treatment effects across sites*, input 0.4; and in the field of *Level-one error variance*, input 2.25. The field for *Power* is left blank because it will be calculated. The default *Significance level* 0.05 is used. The default *H1* is a two-sided test. In the field of *Type of analysis*, select "Test treatment main effect". By clicking the "**Calculate**" button, the statistical power is 0.8067.

The input and output for calculating power for testing the difference between the two treatments are given in Figure 13.4.3. In the field of *Sample size*, input 30; in the field of *Effect size of two treatment difference*, input 0.2; in the field of *Number of clusters*, input 20; in the field of *Variance of treatment effects across sites*, input 0.4; and in the field of *Level-one error variance*, input 2.25. The field for *Power* is left blank because it will be calculated. The default *Significance level* 0.05 is used. The default *H1* is a two-sided test. In the field of *Type of analysis*, select "Compare two treatments". By clicking the "**Calculate**" button, the statistical power is 0.2071.

The input and output for calculating power for the omnibus test is given in Figure 13.4.4. In the field of *Sample size*, input 30; in the field of *Number of clusters*, input 20; in the field of *Effect size of treatment main effect*, input 0.43; in the field of *Effect size of two treatment difference*, input 0.2; in the field of *Number of clusters*, input 20; in the field of *Variance of treatment effects across sites*, input 0.4; and in the field of *Level-one error variance*, input 2.25. The field for *Power* is left blank because it will be calculated. The default *Significance level* 0.05 is used. The default *H1* is a two-sided test. In the field of *Type of analysis*, select "Ominibus test". By clicking the "**Calculate**" button, the statistical power is 0.7951.

Multisite randomized trials with 3 arms

Figure 13.4.2: Input and output for calculating power for testing the treatment main effect using MRT with 3 arms in Example 13.4.1

Parameters (Help)

Sample size	30
Effect size of treatment main effect (Calculator)	0.43
Effect size of two treatment difference	
Number of clusters	20
Variance of treatment effects	0.4
Level 1 error variance	2.25
Power	
Significance level	0.05
Power curve	No power curve
H1	Two-sided test
Type of analysis	Main effect
Note	Multisite randomized tri

Calculate

Output

```
Multisite randomized trials with 3 arms

    J  n   f1 tau  sg2  power alpha
   20 30 0.43 0.4 2.25 0.8067  0.05

NOTE: n is the number of subjects per cluster
URL: http://psychstat.org/mrt3arm
```

The input and output for plotting a power curve for testing the difference between the two treatments in Example 13.4.1 are given in Figure 13.4.5. The cluster size ranges from 20 to 120 with an interval of 20. In the field of *Number of clusters*, the input is 20:120:20. In the field of *Type of analysis*, select "Compare two treatments". We also choose "*Show power curve*" from the drop-down menu of *Power curve*. In the output, the power for each site size from 20 to 120 with the interval 20 is listed. The power curve is displayed at the bottom of the output as shown in Figure 13.4.6. From the power curve, to get a power of 0.8, more than 110 clinics are required if each clinic has 30 patients. Similarly, one can obtain a power curve for different sample sizes by

Example 13.4.2: Power curve with different sample sizes or cluster sizes

Multisite randomized trials with 3 arms

Parameters (Help)

Sample size	30
Effect size of treatment main effect (Calculator)	
Effect size of two treatment difference	0.2
Number of clusters	20
Variance of treatment effects	0.4
Level 1 error variance	2.25
Power	
Significance level	0.05
Power curve	No power curve ⇕
H1	Two-sided test ⇕
Type of analysis	Treatment comparison ⇕
Note	Multisite randomized tri

Calculate

Output

```
Multisite randomized trials with 3 arms

    J  n  f2 tau  sg2  power alpha
   20 30 0.2 0.4 2.25 0.2071  0.05


NOTE: n is the number of subjects per cluster
URL: http://psychstat.org/mrt3arm
```

specifying a sequence of numbers in the field *Sample size*. One can also obtain power curves with different sample sizes or cluster sizes for the other two tests (testing treatment main effect and omnibus test) in a similar way.

In practice, a power of 0.8 is often desired. For the test of comparing the two treatments, given the power, the cluster size can be calculated as shown in Figure 13.4.7. In this situation, the field of *Number of clusters* is left blank while the input for the *Power* field is 0.8. In the output, we can see 111 clinics are needed to obtain a power of 0.8 if each clinic has 30 patients. One can also obtain the required sample size by leaving *Sample size* field blank and specifying a cluster size in

Example 13.4.3: Calculate sample size (or cluster size) given power and effect size

Multisite randomized trials with 3 arms

Figure 13.4.4: Input and output for calculating power for the omnibus test using MRT with 3 arms in Example 13.4.1

Parameters (Help)

Sample size	30
Effect size of treatment main effect (Calculator)	0.43
Effect size of two treatment difference	0.2
Number of clusters	20
Variance of treatment effects	0.4
Level 1 error variance	2.25
Power	
Significance level	0.05
Power curve	No power curve ⬍
H1	Two-sided test ⬍
Type of analysis	Omnibus test ⬍
Note	Multisite randomized tri:

Calculate

Output

```
Multisite randomized trials with 3 arms

    J   n    f1   f2 tau   sg2  power alpha
   20  30 0.43  0.2 0.4  2.25 0.7951  0.05


NOTE: n is the number of subjects per cluster
URL: http://psychstat.org/mrt3arm
```

the field of *Number of clusters*.

13.5 Use R Package WebPower

The power calculation for MRT with three arms is conducted using the R function wp.mrt3arm function. The detail of the function is:

```
wp.mrt3arm(n=NULL, f1=NULL, f2=NULL, J=NULL, tau=NULL, sg2=NULL,
    power=NULL, alpha=0.05, alternative = c("two.sided", "one.
    sided"), type=c("main","treatment","omnibus"))
```

The R input and output for the examples used in the previous section are given below:

n: sample size
f1: effect size for treatment main effect
f2: effect size for the difference between two treatments
J: number of clusters/sites
tau: variance of treatment effects across sites
sg2: level-one error variance
alpha: significance level
power: statistical power
alternative: two-sided or one-sided analysis
type: main: the difference between the average treatment arms and the control arm; treatment: the difference between the two treatment arms; omnibus: the three arms are all equivalent.

Multisite randomized trials with 3 arms

Parameters (Help)

Sample size	30
Effect size of treatment main effect (Calculator)	
Effect size of two treatment difference	0.2
Number of clusters	20:120:20
Variance of treatment effects	0.4
Level 1 error variance	2.25
Power	
Significance level	0.05
Power curve	Show power curve ⬍
H1	Two-sided test ⬍
Type of analysis	Treatment comparison ⬍
Note	Multisite randomized tria

Calculate

Figure 13.4.5: Input and output for power curve for testing the difference between two treatments using MRT with 3 arms in Example 13.4.2

Output

```
Multisite randomized trials with 3 arms

     J  n  f2 tau  sg2  power alpha
    20 30 0.2 0.4 2.25 0.2071  0.05
    40 30 0.2 0.4 2.25 0.3805  0.05
    60 30 0.2 0.4 2.25 0.5337  0.05
    80 30 0.2 0.4 2.25 0.6594  0.05
   100 30 0.2 0.4 2.25 0.7572  0.05
   120 30 0.2 0.4 2.25 0.8304  0.05

NOTE: n is the number of subjects per cluster
URL: http://psychstat.org/mrt3arm
```

```
> ## calculate power for main effect given sample size and effect
    size
> wp.mrt3arm(n=30, f1=0.43, J=20, tau=.4, sg2=2.25, power=NULL)

Multilevel model multisite randomized trials with three arms

    J   n   f1 tau  sg2     power alpha
   20 30 0.43 0.4 2.25 0.8066964  0.05

NOTE: n is the number of observations in each cluster

WebPower URL: http://psychstat.org/mrt3arm

> ## for testing differences between treatment effects
> wp.mrt3arm(n=30, f2=0.2, J=20, tau=.4, sg2=2.25, power=NULL,
    type="treatment")
```

Figure 13.4.6: Power curve for testing the difference between two treatments using MRT with 3 arms in Example 13.4.2

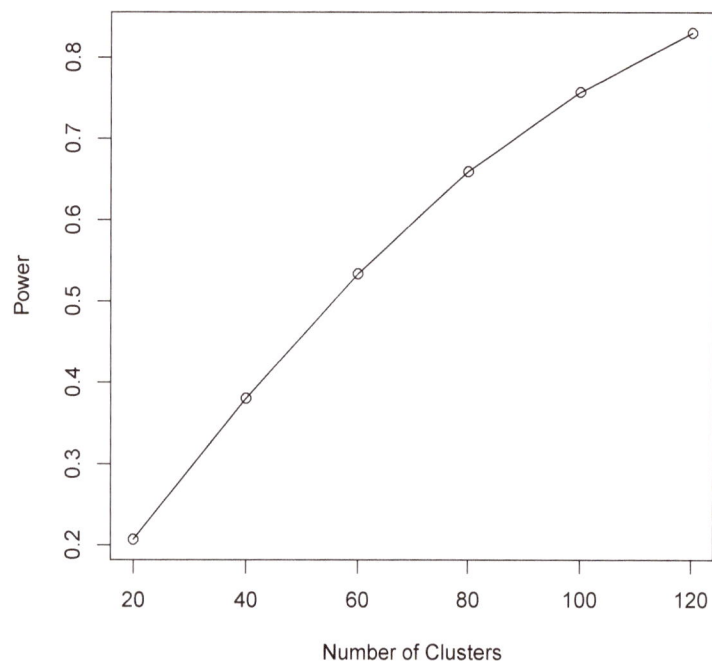

```
Multilevel model multisite randomized trials with three arms

    J  n  f2 tau  sg2     power alpha
   20 30 0.2 0.4 2.25 0.2070712  0.05


NOTE: n is the number of observations in each cluster

WebPower URL: http://psychstat.org/mrt3arm

> ## for testing site variability
> wp.mrt3arm(n=30, f1=.43, f2=0.2, J=20, tau=.4, sg2=2.25, power=
    NULL, type="omnibus")

Multilevel model multisite randomized trials with three arms

    J  n   f1  f2 tau  sg2     power alpha
   20 30 0.43 0.2 0.4 2.25 0.7950757  0.05


NOTE: n is the number of observations in each cluster

WebPower URL: http://psychstat.org/mrt3arm

>
> ## power curve
```

Multisite randomized trials with 3 arms

Parameters (Help)

Sample size

Effect size of treatment main effect (Calculator)	0.43
Effect size of two treatment difference	
Number of clusters	20
Variance of treatment effects	0.4
Level 1 error variance	2.25
Power	0.8
Significance level	0.05
Power curve	No power curve ⬍
H1	Two-sided test ⬍
Type of analysis	Main effect ⬍
Note	Multisite randomized tri:

Calculate

Figure 13.4.7: Input and output for sample size planning for MRT with 3 arms in Example 13.4.3

Output

```
Multisite randomized trials with 3 arms

   J     n    f1 tau  sg2 power alpha
  20 28.62 0.43 0.4 2.25   0.8  0.05


NOTE: n is the number of subjects per cluster
URL: http://psychstat.org/mrt3arm
```

```
> res <- wp.mrt3arm(n=30, f2=0.2, J=seq(20,120,10), tau=.4, sg2
    =2.25, power=NULL, type="treatment")
> res

Multilevel model multisite randomized trials with three arms

    J  n  f2 tau  sg2       power alpha
   20 30 0.2 0.4 2.25 0.2070712  0.05
   30 30 0.2 0.4 2.25 0.2953799  0.05
   40 30 0.2 0.4 2.25 0.3804554  0.05
   50 30 0.2 0.4 2.25 0.4603091  0.05
   60 30 0.2 0.4 2.25 0.5337417  0.05
   70 30 0.2 0.4 2.25 0.6001544  0.05
```

```
 80 30 0.2 0.4 2.25 0.6593902  0.05
 90 30 0.2 0.4 2.25 0.7116052  0.05
100 30 0.2 0.4 2.25 0.7571648  0.05
110 30 0.2 0.4 2.25 0.7965644  0.05
120 30 0.2 0.4 2.25 0.8303690  0.05


NOTE: n is the number of observations in each cluster


WebPower URL: http://psychstat.org/mrt3arm


> plot(res,'J','power')  ## generate power curve
>
> ## sample size given effect size, sample size, power
> wp.mrt3arm(n=NULL, f1=0.43, J=20, tau=.4, sg2=2.25, power=0.8)


Multilevel model multisite randomized trials with three arms


    J        n   f1 tau  sg2 power alpha
   20 28.61907 0.43 0.4 2.25   0.8  0.05


NOTE: n is the number of observations in each cluster


WebPower URL: http://psychstat.org/mrt3arm
```

13.6   *Effect Size for MRT with 3 Arms*

As mentioned previously, there are three types of tests that may be of interest in a 3-arm MRT: (1) testing the treatment main effect, (2) testing the difference between two treatments, and (3) the omnibus test. Given the mean difference between the treatment 1 and control (μ_{D1}), mean difference between the treatment 2 and control (μ_{D2}), and the level 1 error variance (σ^2), the effect sizes of the first two tests can be calculated as

$$f_1 = \frac{(\mu_{D1} + \mu_{D2})/2}{\sqrt{\sigma^2}}, \tag{13.6.1}$$

$$f_2 = \frac{\mu_{D1} - \mu_{D2}}{\sqrt{\sigma^2}}. \tag{13.6.2}$$

Note that effect size is not defined under the omnibus test. To do power estimation and sample size planning for the omnibus test, one needs to specify the effect sizes for testing treatment main effect and testing the difference between two treatments. The specification of the effect sizes can be assisted by an online calculator. In the interface in Figure 13.4.1, clicking the link "Calculator" brings up the calculator. The calculator, as shown in Figure 13.6.1, calculates the effect sizes directly based on equations 13.6.1 and 13.6.2.

Effect Size Calculator for Two Level MRT with 3 Arms
1. Effect size parameter input

Mean difference between treatment 1 and control   0.5

Mean difference between treatment 2 and control   0.8

Level 1 error variance   2.25

Calculate

Effect size output

The effect size for treatment main effect f1 = **0.4333**
The effect size for comparing the two treatments f2 = **0.2000**

Consider the problem in Example 13.4.1, the data can be input as in Figure 13.6.1. At the bottom, the effect sizes for the two tests are presented.

Alternatively,

one can upload a set of empirical data to estimate the effect size. Only registered users can use this method to protect data privacy. The data have to be in the plain text format where the first column of the data is the ID variable, the second column represents the cluster, the third column is the outcome variable and the fourth column is the condition variable with 0 being the control condition, 1 being the first treatment and 2 being the second treatment. The first line of the data should be the variable names. Figure 13.6.2 shows the use of the data in http://psychstat.org/mrt3data and the output – including the estimated effect sizes and the results from conducting the hypothesis testing.

Example data:

```
ID cluster score group
1      1      1     0
2      1      2     0
3      1      6     1
4      1      8     1
5      1      4     2
6      1      5     2
```

13.7   *Technical Details*

In this section, we present the models for 2-arm MRT and 3-arm MRT, and formulas for power estimation under the two models. Details leading to the power calculation can be found in Raudenbush & Liu (2000) and Liu (2013).

The model for a 2-arm MRT can be expressed as

$$Y_{ij} = \beta_{0j} + \beta_{1j}X_{ij} + e_{ij},$$

$$\beta_{0j} = \gamma_{00} + u_{0j}, \ \beta_{1j} = \gamma_{10} + u_{1j}.$$

$$e_{ij} \sim N(0,\sigma^2), \ \begin{pmatrix} u_{0j} \\ u_{1j} \end{pmatrix} \sim N\left(\mathbf{0}, \begin{bmatrix} \tau_{00} & \tau_{01} \\ \tau_{10} & \tau_{11} \end{bmatrix}\right),$$

Effect Size Calculator for Two Level MRT with 3 Arms
2. From empirical data analysis

Figure 13.6.2: Effect size calculation based on an empirical set of data

Upload data file:

Choose File | MRT3.txt   Calculate

Effect size output

The effect size of treatment main effect f1 = **-0.6214**
The effect size of comparing the two treatments f2 = **-0.3551**

Test of fixed effects (X: Treatment main effect; X1-X2: Comparing

```
              Estimate  Std. Error   t value     p.value
(Intercept)   2.166667   0.2861225  7.5725153  0.004776785
X            -0.875000   0.4874111 -1.7951992  0.170497896
X1-X2        -0.500000   0.5618296 -0.8899496  0.439053442
```

where

- Y_{ij} is the ith outcome in the jth cluster ($i = 1, 2, ..., N$; $j = 1, 2, ...J$),

- X_{ij} is the treatment indicator of subject i at site j (0.5 for treatment and -0.5 for control),

- β_{0j} is the mean of the outcome at the jth site,

- β_{1j} is the mean difference between treatment and control at the jth site,

- γ_{00} is the grand mean,

- γ_{10} is the grand mean difference between the treatment and control across all the sites (treatment main effect),

- the level-one residual variance σ^2 represents level-one error variance,

- and the level-2 residual variances τ_{00} and τ_{11} represent site variability and variance of the mean differences among sites, respectively.

Power for testing the treatment main effect $H_0 : \gamma_{10} = 0$ is calculated by

$$\text{Power} = \begin{cases} 1 - P[T_{J-1,\lambda} < t_0] + P[T_{J-1,\lambda} \leq -t_0] & \text{two} - \text{sided test}, \\ 1 - P[T_{J-1,\lambda} < t_0] & \text{one} - \text{sided test}, \end{cases}$$

(13.7.1)

where

$$\lambda = \frac{\gamma_{10}}{\sqrt{(\frac{4\sigma^2}{n} + \tau_{11})/J}},$$

(13.7.2)

t_0 is the $100(1 - \frac{\alpha}{2})$th percentile for a two-sided test and the $100(1 - \alpha)$th percentile for a one-sided test of the t distribution with $J - 1$ degrees of freedom, and α is the significance level. Power for testing the site variability $H_0 : \tau_{00} = 0$ is

$$\text{Power} = P(F_{J-1,J(n-2)} \geq F_0 \frac{\sigma^2}{n\tau_{00} + \sigma^2}), \qquad (13.7.3)$$

and for testing the variance of treatment effects across sites $H_0 : \tau_{11} = 0$

$$\text{Power} = P(F_{J-1,J(n-2)} \geq F_0 \frac{\sigma^2}{n\tau_{11}/4 + \sigma^2}), \qquad (13.7.4)$$

where F_0 is the $100(1 - \alpha)$th percentile of the F distribution with degrees of freedom $J - 1$ and $J(n - 2)$.

The model for a 3-arm MRT can be expressed as

$$Y_{ij} = \beta_{0j} + \beta_{1j} X_{1ij} + \beta_{2j} X_{2ij} + e_{ij},$$

$$\beta_{0j} = \gamma_{00} + u_{0j}, \ \beta_{1j} = \gamma_{10} + u_{1j}, \ \beta_{2j} = \gamma_{20} + u_{2j},$$

$$Var(e_{ij}) = \sigma^2, \ Var(u_{0j}) = \tau_{00}, \ Var(u_{1j}) = \tau_{11}, \ Var(u_{2j}) = \tau_{22},$$

where

- Y_{ij} is the ith outcome in jth cluster ($i = 1, 2, ..., N$; $j = 1, 2, ...J$),

- X_{1ij} is used to compare the average outcome of the two treatment arms with that of the control arm (1/3 for the first treatment, 1/3 for the second treatment and -2/3 for the control condition),

- X_{2ij} is used to contrast the average outcome between the two treatment arms (1/2 for the first treatment, -1/2 for the second treatment and 0 for the control condition),

- β_{0j} is mean of the jth site,

- β_{1j} is the mean difference between average treatment and control of the jth site,

- β_{2j} is the mean difference between the two treatments of the jth site,

- γ_{00} is grand mean,

- γ_{10} is the contrast between average of the two treatments and control,

- and γ_{20} is the contrast between the two treatments.

Power for testing the treatment main effect $H_0 : \gamma_{10} = 0$ is calculated by

$$\text{Power} = \begin{cases} 1 - P[T_{J-1,\lambda_1} < t_0] + P[T_{J-1,\lambda_1} \leq -t_0] & \text{two} - \text{sided test,} \\ 1 - P[T_{J-1,\lambda_1} < t_0] & \text{one} - \text{sided test,} \end{cases}$$

(13.7.5)

where

$$\lambda_1 = \frac{\gamma_{10}}{\sqrt{(4.5\frac{\sigma^2}{n} + \tau_{11})/J}},$$

(13.7.6)

t_0 is the $100(1 - \frac{\alpha}{2})$th percentile for a two-sided test and the $100(1 - \alpha)$th percentile for a one-sided test of the t distribution with $J - 1$ degrees of freedom, and α is the significance level. Power for comparing the two treatments $H_0 : \gamma_{20} = 0$ is calculated by

$$\text{Power} = \begin{cases} 1 - P[T_{J-1,\lambda_2} < t_0] + P[T_{J-1,\lambda_2} \leq -t_0] & \text{two} - \text{sided test,} \\ 1 - P[T_{J-1,\lambda_2} < t_0] & \text{one} - \text{sided test,} \end{cases}$$

(13.7.7)

where

$$\lambda_2 = \frac{\gamma_{20}}{\sqrt{(6\frac{\sigma^2}{n} + \tau_{22})/J}},$$

(13.7.8)

t_0 is the $100(1 - \frac{\alpha}{2})$th percentile for a two-sided test and the $100(1 - \alpha)$th percentile for a one-sided test of the t distribution with $J - 1$ degrees of freedom, and α is the significance level. Power for the omnibus test $H_0 : \gamma_{10} = \gamma_{20} = 0$ is calculated by

$$\text{Power} = P(F_{2,2(J-1),\lambda_3} \geq F_0),$$

(13.7.9)

where

$$\lambda_3 = \lambda_1^2 + \lambda_2^2 = \frac{\gamma_{10}}{\sqrt{(4.5\frac{\sigma^2}{n} + \tau_{11})/J}} + \frac{\gamma_{20}}{\sqrt{(6\frac{\sigma^2}{n} + \tau_{22})/J}},$$

(13.7.10)

F_0 is the $100(1 - \alpha)$th percentile of the F distribution with degrees of freedom 2 and $2(J - 1)$.

13.8 Exercises

1. A researcher plans to design a multisite randomized trial for studying the effects of aerobic exercise (specifically walking) on cognitive improvement in Alzheimer's disease. The data will be collected from community-dwelling adults who have moderate Alzheimer's disease in 5 communities. To determine the sample size at each community, the researcher has found in the literature that walking for 2 months

increased scores on a test of cognitive abilities by 2.5. Further, the variance of the treatment effects across communities is estimated to be 2 and the person-specific variance is estimated to be 1. How many subjects are needed in each community to get a power 0.8 at the alpha level 0.05?

2. Using the same information in Exercise 1, generate a power curve with the sample size ranging from 10 to 100 with an interval of 10. What would be the power when the sample size is 60?

14 *Statistical Power Analysis for Simple Mediation via Sobel Test*

Zhiyong Zhang
Department of Psychology
University of Notre Dame

Mediation models are widely used in social and behavioral sciences as demonstrated in recent books by Hayes (2013) and MacKinnon (2008). Mediation models are useful because they can be used to investigate the underlying mechanisms related to why an input variable influences an output variable. We illustrate how to use WebPower to conduct statistical power analysis for a simple mediation model based on the Sobel test.

Figure 14.0.1 displays the path diagram of a simple mediation model. In the figure, x, m, and y represent the input variable, the mediation variable, and the outcome variable, respectively. In this model, the total effect of x on y, c'+a*b, consists of the direct effect c' and the mediation effect θ=a*b, the multiplication of the direct effect of x on m and the direct effect of m on y. The mediation effect is also called the indirect effect because it is the effect of x on y indirectly through m.

Statistical power analysis for mediation can be viewed as concerning a test whether the mediation effect ($\theta = ab$) is significantly different from 0. More specifically, we have the null and alternative hypotheses

$$H_0 : \theta = 0 \text{ vs. } H_1 : \theta \neq 0.$$

To test the mediation effect, the Sobel test has been used in which the test statistic

$$z = \frac{\hat{a}\hat{b}}{se(\hat{a}\hat{b})}$$

is assumed to follow a normal distribution. Based on the Sobel test, statistical power analysis can be conducted.

Figure 14.0.1: Path diagram of a simple mediation model.

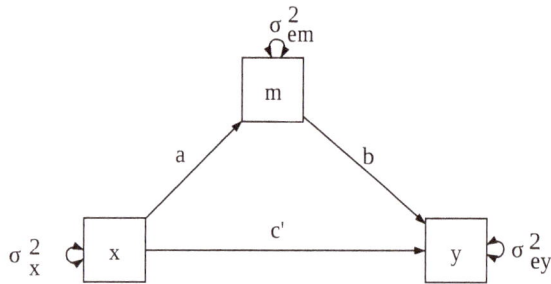

14.1 How to Conduct Power Analysis for Mediation Analysis

The primary software interface for power analysis for mediation analysis is shown in Figure 14.1.1. Within the interface, a user can supply different parameter values and select different options for power analysis.

http://psychstat.org/mediation

- *Sample size.* The total number of participants. Multiple sample sizes can be provided in two ways to calculate power for each sample size. First, multiple sample sizes can be supplied and separated by white spaces, e.g., 100 150 200 will calculate power for the three sample sizes 100, 150 and 200. A sequence of sample sizes can be generated using the method s:e:i with s denoting the starting sample size, e as the ending sample size, and i as the interval. Note that the values are separated by colon ":". For example, 100:150:10 will generate a sequence of sample sizes - 100 110 120 130 140 150. The default sample size, as shown in Figure 14.1.1, is 100.

- *Path a.* The coefficient from the input variable x to the mediator m.

- *Path b.* The coefficient from the mediator variable m to the outcome variable y.

- *Variance of x:* σ_x^2. The variance of the input variable x.

- *Variance of m:* σ_m^2. The variance of the mediation variable m. Note that $\sigma_m^2 = \sigma_{em}^2 + a^2\sigma_x^2$.

- *Error variance of y:* σ_{ey}^2. The error or residual variance of the output variable y.

- *Significance level.* The type I error rate is used for power calculation, and it is usually set at the default value 0.05.

- *Power.* The desired statistical power can be specified here.

- *Power curve.* One can also request the plot of a power curve if multiple sample sizes or effect sizes are provided.

- *Note.* A note (less than 200 characters) can also be provided to provide basic information on the analysis for future reference for registered users.

Among the following input, *Sample size*, *Path a*, *Path b*, *Variance of x*, *Variance of m*, *Error variance of y*, *Significance level*, and *Power*, one and only one can be left blank. However, a solution is not guaranteed when calculating a quantity other than power. The path coefficients are combined to form the effect size of interest.

14.1.1   *Examples*

Simple Mediation via Sobel Test

Figure 14.1.1: Software interface of power analysis for simple mediation analysis

Parameters (Help)

Sample size	100
Path a	0.5
Path b	0.5
Variance of x	1
Variance of m	1
Error variance of y	1
Significance level	0.05
Power	
Power curve	No power curve ↕
Note	Sobel test

Calculate

Suppose we want to investigate whether home environment (m) is a mediator between the relationship of mother's education (x) and child's mathematical ability (y). Furthermore, we know $a = b = 0.5$, $\sigma_x^2 = \sigma_{em}^2 = \sigma_{ey}^2 = 1$. Then, we want to know the statistical power we can achieve with a sample of 100 participants at the significance level 0.05.

Example 14.1.1: Calculate power given sample size and model parameter values

The input and output for calculating power for this study are given in Figure 14.1.2. The parameter values specified above are input in each field. The field for *Power* is left blank because it will be calculated. By clicking the "**Calculate**" button, the statistical power is given in the output. For the current design, the power is 0.9337. Therefore, a study with a sample size 100 would achieve sufficient power.

Simple Mediation via Sobel Test

Parameters (Help)

Sample size	100
Path a	0.5
Path b	0.5
Variance of x	1
Variance of m	1
Error variance of y	1
Significance level	0.05
Power	
Power curve	No power curve ↕
Note	Sobel test

Calculate

Output

```
Power for simple mediation

  n  power    a   b varx varm vary alpha
100 0.9337 0.5 0.5    1    1    1  0.05


URL: http://psychstat.org/mediation
```

Example 14.1.2: Power curve

A power curve is a line plot of statistical power along with the given sample sizes. In Example 14.1.1, the power is 0.9337 with the sample size 100. What is the power for a different sample size, say, 50? One can investigate the power of different sample sizes and plot a power curve.

The input and output for calculating power for the study in Example 14.1.1 with a sample size from 50 to 100 with an interval of 10 are given in Figure 14.1.3. Note that in the *Sample size* field, the input is 50:100:10. In the output, the power for each sample size from 50 to 100 with the interval 10 is listed. Especially, with the sample size 70, the power is about 0.826. In the input, we also choose "*Show power curve*" for the

Power curve parameter. In the output, the power curve is displayed at the bottom of the output as shown in Figure 14.1.4. The power curve can be used for interpolation. For example, to get a power 0.8, a sample size about 65 is needed.

Simple Mediation via Sobel Test

Figure 14.1.3: Input and output for obtaining power curve for simple mediation analysis in Example 14.1.2

Parameters (Help)

Sample size	50:100:10
Path a	0.5
Path b	0.5
Variance of x	1
Variance of m	1
Error variance of y	1
Significance level	0.05
Power	
Power curve	Show power curve $\updownarrow$
Note	Sobel test

Calculate

Output

```
Power for simple mediation

      n   power     a    b varx varm vary alpha
     50  0.6878   0.5  0.5    1    1    1  0.05
     60  0.7653   0.5  0.5    1    1    1  0.05
     70  0.8260   0.5  0.5    1    1    1  0.05
     80  0.8725   0.5  0.5    1    1    1  0.05
     90  0.9076   0.5  0.5    1    1    1  0.05
    100  0.9337   0.5  0.5    1    1    1  0.05


URL: http://psychstat.org/mediation
```

In practice, a power 0.8 or higher is often desired. Given the power, the sample size can also be calculated as shown in Figure 14.1.5. In this situation, the *Sample size* field is left blank while in the *Power* field, the value 0.9 is input. In the output, we can see that a sample size 88 is needed to obtain a power 0.9.

Example 14.1.3: Calculate sample size given power and effect size

In studying the mediation effect, the researcher might not have

Example 14.1.4: Calculate an effect given power and sample size

Figure 14.1.4: Power curve for simple mediation analysis in Example 14.1.2

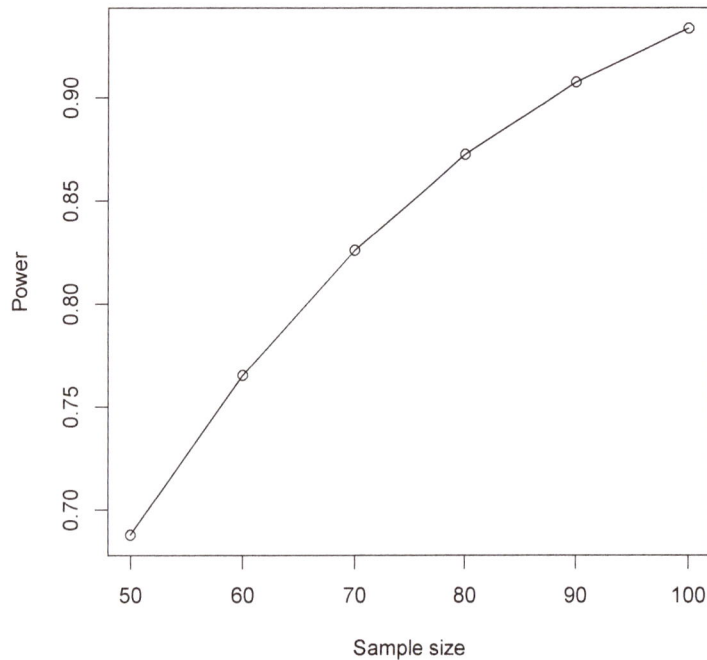

enough information to determine the size of an effect. In this case, the researcher can estimate the minimum effect to achieve certain power given a sample size. Such an effect can help the researcher determine whether it is meaningful to conduct a study. For example, if a very high effect is needed, it might be difficult to achieve significant results in practical data collection anyway. In Figure 14.1.6, we calculate the path a for the mediation model by proving both sample size and power as well as other parameters. Note that the path a has to be at least 0.734 to have a power 0.9 with the sample size 100.

14.2 Using WebPower for Power Analysis for Simple Mediation

The power calculation for the simple mediation analysis is conducted using the R function wp.mediation. The detail of the function is:

```
wp.mediation(n=NULL, power=NULL, a=.5, b=.5, varx=1, vary=1, varm
    =1, alpha = 0.05)
```

n: sample size
power: statistical power
a: coefficient from x to m
b: coefficient from m to y
varx: variance of x
varm: variance of y
vary: error variance for y
alpha: significance level

The R input and output for the examples discussed in the previous section are given below.

```
> wp.mediation(n=100, a=.5, b=.5, varx=1, varm=1, vary=1)
```

Simple Mediation via Sobel Test

Figure 14.1.5: Input and output for obtaining sample size for simple mediation analysis in Example 14.1.3

Parameters (Help)

Sample size

Path a   0.5

Path b   0.5

Variance of x   1

Variance of m   1

Error variance of y 1

Significance level 0.05

Power   0.9

Power curve   No power curve   ⬍

Note   Sobel test

Calculate

Output

```
Power for simple mediation

      n power    a   b varx varm vary alpha
  87.56   0.9 0.5 0.5    1    1    1  0.05


URL: http://psychstat.org/mediation
```

```
Power calculation for simple mediation based on Sobel test

    n      power   a   b varx varm vary alpha
  100 0.9337271 0.5 0.5    1    1    1  0.05

WebPower URL: http://psychstat.org/mediation

> wp.mediation(n=NULL, power=.9, a=.5, b=.5, varx=1, varm=1, vary
    =1)

Power calculation for simple mediation based on Sobel test

        n power   a   b varx varm vary alpha
  87.56182   0.9 0.5 0.5    1    1    1  0.05


WebPower URL: http://psychstat.org/mediation

>
```

Figure 14.1.5: Input and output for obtaining sample size for simple mediation analysis in Example 14.1.3

Simple Mediation via Sobel Test

Parameters (Help)

Sample size	100
Path a	
Path b	0.5
Variance of x	1
Variance of m	1
Error variance of y	1
Significance level	0.05
Power	0.9
Power curve	No power curve ⬍
Note	Sobel test

Calculate

Figure 14.1.6: Input and output for obtaining a minimum effect for simple mediation analysis in Example 14.1.3

Output

```
Power for simple mediation

     n power       a    b varx varm vary alpha
   100   0.9 0.7335 0.5    1    1    1  0.05


URL: http://psychstat.org/mediation
```

```
> ## power curve
> res <- wp.mediation(n=seq(50,100,5), a=.5, b=.5, varx=1, varm
    =1, vary=1)
> res

Power calculation for simple mediation based on Sobel test

     n     power   a   b varx varm vary alpha
    50 0.6877704 0.5 0.5    1    1    1  0.05
    55 0.7287681 0.5 0.5    1    1    1  0.05
    60 0.7652593 0.5 0.5    1    1    1  0.05
    65 0.7975459 0.5 0.5    1    1    1  0.05
    70 0.8259584 0.5 0.5    1    1    1  0.05
    75 0.8508388 0.5 0.5    1    1    1  0.05
    80 0.8725282 0.5 0.5    1    1    1  0.05
    85 0.8913577 0.5 0.5    1    1    1  0.05
    90 0.9076417 0.5 0.5    1    1    1  0.05
    95 0.9216744 0.5 0.5    1    1    1  0.05
   100 0.9337271 0.5 0.5    1    1    1  0.05
```

```
WebPower URL: http://psychstat.org/mediation

> plot(res)  ## generate power curve
>
> wp.mediation(n=100, power=.9, a=NULL, b=.5, varx=1, varm=1,
    vary=1)


Power calculation for simple mediation based on Sobel test

     n power         a   b varx varm vary alpha
   100   0.9 0.7335197 0.5    1    1    1  0.05


WebPower URL: http://psychstat.org/mediation
```

14.3   *Technical Details*

Consider a simple mediation model

$$m_i = a_0 + a * x_i + e_{m_i}$$
$$y_i = b_0 + b * m_i + c * x_i + e_{y_i}$$

where $e_{m_i} \sim N(0, \sigma_{e_m}^2)$ and $e_{y_i} \sim N(0, \sigma_{e_y}^2)$. The mediation effect is $ab = a * b$.

The Sobel test statistic is

$$Z = \frac{\hat{a}\hat{b}}{\hat{\sigma}_{ab}}$$

where $\hat{\sigma}_{ab}^2 = \hat{a}^2 * \hat{\sigma}_b^2 + \hat{b}^2 * \hat{\sigma}_a^2$. From regression analysis, we have

$$\hat{\sigma}_a^2 = \frac{\sigma_{e_m}^2}{n\sigma_x^2}$$
$$\hat{\sigma}_b^2 = \frac{\sigma_{e_y}^2}{n\sigma_m^2(1 - \rho_{xm}^2)}$$

where σ_x^2 and σ_m^2 are variance for x and m and ρ_{xm} is the correlation between x and m.

Furthermore, because $\hat{a} = \rho_{xm} * \sigma_m / \sigma_x$, we have

$$\rho_{xm} = \hat{a}\sigma_x/\sigma_m$$
$$\sigma_{em}^2 = \sigma_m^2(1 - \rho_{xm}^2) = \sigma_m^2 - \hat{a}^2\sigma_x^2.$$

Then

$$\hat{\sigma}_a^2 = \frac{\sigma_m^2 - a^2\sigma_x^2}{n\sigma_x^2},$$
$$\hat{\sigma}_b^2 = \frac{\sigma_{ey}^2}{n(\sigma_m^2 - a^2\sigma_x^2)}.$$

Therefore, the Sobel test depends on the sample size, the coefficients a and b, the variances of x and m, and the residual variance of y denoted by $\hat{\sigma}_{e_y}^2$ as in

$$Z = \frac{\hat{a}\hat{b}}{\sqrt{\hat{a}^2 * \frac{\sigma_{e_y}^2}{n(\sigma_m^2 - a^2\sigma_x^2)} + \hat{b}^2 * \frac{\sigma_m^2 - a^2\sigma_x^2}{n\sigma_x^2}}}$$

To calculate power using WebPower, one needs to provide information on (1) sample size, (2) coefficient a, (3) coefficient b, (4) variance of x (σ_x^2), (5) variance of m (σ_m^2), (6) error variance for y ($\sigma_{e_y}^2$), and (7) the significance level α. If the power is provided, the needed sample size can also be calculated.

14.4 Exercises

1. A study found that processing speed mediated the relationship between age and everyday living in a negative but insignificant way. A researcher believes a study with a larger sample size would lead to significant results. From the current study, he decides the population parameter values should be $a = -0.3, b = 0.4, \sigma_x^2 = \sigma_m^2 = \sigma_{e_y}^2 = 1$. What is the required sample size to achieve a power 0.8?

2. Suppose in Example 1, the researcher has resources to collect data from 40 participants. What would be his power?

15 *Statistical Power Analysis for Structural Equation Modeling*

Zhiyong Zhang
Department of Psychology
University of Notre Dame

Structural equation modeling (SEM) is one of the most widely used methods in social and behavioral sciences. SEM is a multivariate technique that is used to study relationships between observed and latent variables as well as among observed and latent variables. It can be viewed as a combination of factor analysis and path analysis. Two methods are widely used in power analysis for SEM. The first one is based on the likelihood ratio test proposed by Satorra & Saris (1985) and the second one is based on the root mean square error of approximation (RMSEA) proposed by MacCallum et al. (1996). We will show how to conduct power analysis for both methods using WebPower.

15.1 *Power Analysis Using the Satorra & Saris (1985) Method*

The primary software interface for SEM using the Satorra & Saris (1985) method is shown in Figure 15.1.1. Within the interface, a user can supply different parameter values and select different options for power analysis.

http://psychstat.org/semchisq

- *Sample size* is the total number of participants. Multiple sample sizes can be provided in two ways to calculate power for each sample size. First, multiple sample sizes can be supplied and separated by white spaces, e.g., 100 150 200 will calculate power for the three sample sizes 100, 150 and 200. Second, a sequence of sample sizes can be generated using the method s:e:i with s denoting the starting sample size, e as the ending sample size, and i as the interval. Note that the values are separated by colon ":". For example, 100:150:10 will generate a sequence of sample sizes - 100 110 120 130 140 150. The default sample size is 100.

- The *degrees of freedom* of the chi-squared test. Multiple degrees of freedom can be provided as for the sample size.

- The *Effect size* specifies the population misfit of a SEM model.[1] Multiple effect sizes or a sequence of effect sizes can also be supplied using the same method for sample size. By default, the value is 0.1. Determining the effect size is critical but not trivial work. To help a user obtain effect sizes, a calculator has been developed and can be used by clicking the link "**Calculator**".

- The *Significance level* for power calculation is needed but usually set at the default value 0.05. Multiple significance levels can be supplied by separating them using white spaces.

- The *Power* specifies the desired statistical power. Multiple power values can be supplied by separating them using white spaces.

- In addition to the required input, one can also request the plot of a power curve if multiple sample sizes or effect sizes are provided.

- A note (less than 200 characters) can also be provided to provide basic information on the analysis for future reference for registered users.

[1] The effect size will be discussed in more detail later.

Among the following input, *Sample size, Degrees of freedom, Effect size, Significance level*, and *Power*, one and only one can be left blank. However, note that a solution is not guaranteed when calculating a quantity other than power.

15.2 Examples

Satorra & Saris (1985) presented an example in which Y_1, Y_2, Y_3, Y_4, and X satisfy the following population model

$$
\begin{aligned}
Y_1 &= \gamma_1 X + \zeta_1 \\
Y_2 &= \gamma_2 X + \beta_{21} Y_1 + \zeta_2 \\
Y_3 &= \gamma_3 X + \zeta_3 \\
Y_4 &= \beta_{41} Y_1 + \beta_{42} Y_2 + \beta_{43} Y_3 + \zeta_4
\end{aligned}
\tag{15.2.1}
$$

where $\zeta_i, i = 1, 2, 3, 4$, follows a normal distribution with mean 0 and unknown variance ψ_i, and X has mean 0 and variance 1. The parameter values are

Parameter	γ_1	γ_2	γ_3	β_{41}	β_{42}	β_{43}	β_{21}	ψ_1	ψ_2	ψ_3	ψ_4
value	.4	.5	.4	.4	.4	.4	.2	.84	.61	.84	.27

Suppose a model is fitted without β_{21}. Then, one can investigate the power to detect the path β_{21} through the likelihood ratio test.

Example 15.2.1: Calculate power given sample size and effect size

SEM based on Chi-squared test

Parameters (Help)

Sample size	100
Degrees of freedom	4
Effect size (Calculator)	0.054
Significance level	0.05
Power	
Power curve	No power curve ↕
Note	Satorra & Saris 1985

Calculate

Output

```
Power for SEM (Satorra & Saris, 1985)

   n df effect  power alpha
 100  4  0.054 0.4221  0.05


URL: http://psychstat.org/semchisq
```

The corresponding effect size for $\beta_{21} = 0.2$ is 0.054 and the degrees of freedom for the test is 4.[2] A researcher is interested in the power given that she can collect data from 100 participants.

The input and output for calculating power for this study are given in Figure 15.1.1. In the *Sample size* field, we input 100 and in the *Degrees of freedom* field, we input 4. The *Effect size* is 0.054. The default *Significance level* 0.05 is used. The field for *Power* is left blank because it will be calculated. By clicking the "**Calculate**" button, the statistical power is given in the output. For the current analysis, the power is 0.4221.

[2] The method for calculating the effect size will be explained later.

Satorra & Saris (1985) evaluated the power with significance levels at 0.001, 0.01, 0.025, 0.05 and 0.1. The same analysis can be conducted using WebPower with the input and output given in Figure 15.2.1. Note that in the *Significance level* field, multiple values are specified and separated by white spaces. Power corresponding to each significance level is shown in the output. Clearly, higher significance level leads to higher power.

Example 15.2.2: Calculate power with different significance levels

Satorra & Saris (1985) also used two sample sizes, 100 and 600. WebPower can generate a power curve with different sample sizes. For

Example 15.2.3: Power curve

SEM based on Chi-squared test

Figure 15.2.1: Power for SEM with different significance levels in Example 15.2.2

Parameters (Help)

Sample size	100
Degrees of freedom	4
Effect size (Calculator)	0.054
Significance level	.001 .01 .25 .05 .1
Power	
Power curve	No power curve ⬍
Note	Satorra & Saris 1985

Calculate

Output

```
Power for SEM (Satorra & Saris, 1985)

    n df effect    power alpha
  100  4  0.054 0.06539 0.001
  100  4  0.054 0.20867 0.010
  100  4  0.054 0.74499 0.250
  100  4  0.054 0.42212 0.050
  100  4  0.054 0.55040 0.100


URL: http://psychstat.org/semchisq
```

example, the input and output in Figure 15.2.2 calculate power with a sample size from 100 to 600 with an interval of 100. Note that in the *Sample size* field, the input is 100:600:100. In the output, the power for each sample size from 100 to 600 with the interval 100 is listed. Especially, with the sample size 300, the power is about 0.9146. In the input, we also choose "*Show power curve*" for the *Power curve* field. In the output, the power curve is displayed at the bottom of the output as shown in Figure 15.2.3. The power curve can be used for interpolation. For example, to get a power 0.8, a sample size of about 250 is needed.

In practice, a power 0.8 or higher is often desired. Given the power, the sample size can also be calculated as shown in Figure 15.2.4. In this situation, the *Sample size* field is left blank while in the *Power* field, the value 0.8 is input. In the output, we can see a sample size 231 is needed to obtain a power 0.8.

Example 15.2.4: Calculate sample size given power and effect size

Suppose a researcher does not have enough information to determine

Example 15.2.5: Calculate effect size given power and sample size

SEM based on Chi-squared test

Figure 15.2.2: Input and output for obtaining power curve for SEM in Example 15.2.3

Parameters (Help)

Sample size	100:600:100
Degrees of freedom	4
Effect size (Calculator)	0.054
Significance level	0.05
Power	
Power curve	Show power curve ↕
Note	Satorra & Saris 1985

Calculate

Output

```
Power for SEM (Satorra & Saris, 1985)

     n df effect  power alpha
   100  4  0.054 0.4221  0.05
   200  4  0.054 0.7511  0.05
   300  4  0.054 0.9146  0.05
   400  4  0.054 0.9750  0.05
   500  4  0.054 0.9935  0.05
   600  4  0.054 0.9985  0.05


URL: http://psychstat.org/semchisq
```

the size of effect. In this case, the researcher can find the minimum effect to achieve certain power given the sample size. Such an effect can help the researcher determine whether it is meaningful to conduct a study. For example, if a very large effect is needed, it might be difficult to achieve significant results. In Figure 15.2.5, we calculate the effect size by providing both sample size and power as well as other information. Note that the effect size has to be at least 0.1206 to have a power 0.8 with the sample size 100.

15.3   *Using WebPower for Power Analysis*

The power calculation for the SEM based on the method developed by Satorra & Saris (1985) is conducted using the R function wp.sem.chisq. The detail of the function is:

```
wp.sem.chisq(n = NULL, df = NULL, effect = NULL, power = NULL,
    alpha = 0.05)
```

n: sample size
df: degrees of freedom
effect: effect size
power: statistical power
alpha: significance level

Figure 15.2.3: Input and output for obtaining power curve for SEM in Example 15.2.3

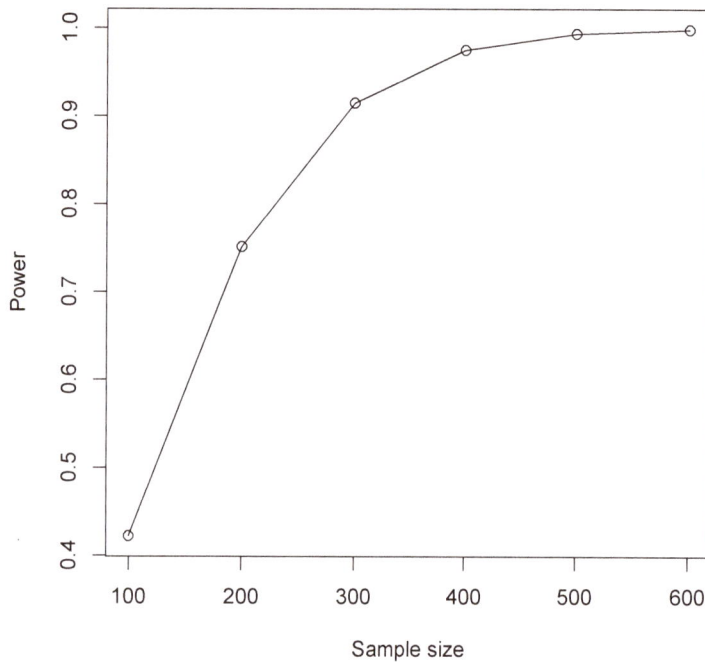

The R input and output for the examples discussed in the previous section are given below

```
> ## calculate power given sample size and effect size
> wp.sem.chisq(n = 100, df = 4, effect = .054, power = NULL,
    alpha = 0.05)

Power analysis for SEM (Satorra & Saris, 1985)

    n df effect     power alpha
  100  4  0.054 0.4221152  0.05


NOTE: Power analysis for SEM (Satorra & Saris, 1985)

WebPower URL: http://psychstat.org/semchisq


>
> ## Power with different alphas
> wp.sem.chisq(n = 100, df = 4, effect = .054, power = NULL,
    alpha = c(.001, .005, .01, .025, .05))

Power analysis for SEM (Satorra & Saris, 1985)

    n df effect     power alpha
  100  4  0.054 0.06539478 0.001
```

SEM based on Chi-squared test

Figure 15.2.4: Input and output for obtaining sample size for SEM in Example 15.2.4

Parameters (Help)

Sample size	
Degrees of freedom	4
Effect size (Calculator)	0.054
Significance level	0.05
Power	0.8
Power curve	No power curve ⬍
Note	Satorra & Saris 1985

Calculate

Output

```
Power for SEM (Satorra & Saris, 1985)

    n df effect power alpha
  222  4  0.054    0.8  0.05


URL: http://psychstat.org/semchisq
```

```
    100  4  0.054 0.14952768 0.005
    100  4  0.054 0.20867087 0.010
    100  4  0.054 0.31584011 0.025
    100  4  0.054 0.42211515 0.050

NOTE: Power analysis for SEM (Satorra & Saris, 1985)

WebPower URL: http://psychstat.org/semchisq

>
> ## power curve
> res <- wp.sem.chisq(n = seq(100,600,100), df = 4, effect =
    .054, power = NULL, alpha = 0.05)
> res

Power analysis for SEM (Satorra & Saris, 1985)

    n df effect     power alpha
  100  4  0.054 0.4221152  0.05
  200  4  0.054 0.7510630  0.05
  300  4  0.054 0.9145660  0.05
  400  4  0.054 0.9750481  0.05
  500  4  0.054 0.9935453  0.05
  600  4  0.054 0.9984820  0.05
```

SEM based on Chi-squared test

Parameters (Help)

Sample size	100
Degrees of freedom	4
Effect size (Calculator)	
Significance level	0.05
Power	0.8
Power curve	No power curve ⬍
Note	Satorra & Saris 1985

Calculate

Output

```
Power for SEM (Satorra & Saris, 1985)

    n df effect power alpha
  100  4 0.1206   0.8  0.05


URL: http://psychstat.org/semchisq
```

```
NOTE: Power analysis for SEM (Satorra & Saris, 1985)

WebPower URL: http://psychstat.org/semchisq

> plot(res)  ## generate power curve
>
> ## calculate the effect size given other information
> wp.sem.chisq(n = 100, df = 4, effect = NULL, power = 0.8, alpha
    = 0.05)

Power analysis for SEM (Satorra & Saris, 1985)

    n df    effect power alpha
  100  4 0.1205597   0.8  0.05

NOTE: Power analysis for SEM (Satorra & Saris, 1985)

WebPower URL: http://psychstat.org/semchisq
```

Figure 15.2.5: Input and output for obtaining a minimum effect size for SEM in Example 15.2.4

Effect Size

The effect size used in this calculator is defined as the difference between two SEM models, a full model M_F and a reduced model M_R. The full model includes all the parameters in the population such as the one in Equation 15.2.1. The reduced model is nested within the full model by setting certain relationship to be null. For example, the reduced model of the full model in Equation 15.2.1 is the one by setting $\beta_{21} = 0$. To get an effect size, one needs to specify a value for each parameter in the full model.

The effect size can be calculated in the following way. First, from the full model, the model implied covariance matrix can be obtained as Σ_F. Then, the reduced model can be fitted to Σ_F. Suppose the estimated covariance matrix for the reduced model is $\hat{\Sigma}_R$. The effect size δ is obtained as

$$\delta = \log |\hat{\Sigma}_R| + \text{tr}(\Sigma_F \hat{\Sigma}_R^{-1}) - \log |\Sigma_F| - p,$$

where p is the number variables for both the full model and reduced model.

A convenient way to get the effect size is to fit the reduced model to Σ_F through SEM software such as lavaan (Rosseel, 2012) with a predefined sample size n. Suppose the obtained chi-squared statistic is λ. Then the effect size is $\delta = \lambda/(n-1)$.

There is an approximate connection between the effect size δ and the RMSEA ϵ where

$$\delta = d\epsilon^2$$

with d denoting the degrees of freedom. MacCallum et al. (1996) have used 0.01, 0.05, and 0.08 to indicate excellent, good, and mediocre fit, respectively. Based on this, we can define small, medium, and large effects as shown in Table 15.4.1. Note that the effect size δ depends on degrees of freedom given the value of RMSEA.

WebPower provides an online calculator to obtain the effect size defined here with two methods.

Method 1: through a predefined population model and parameter values

An example input and output for using this method is given in Figure 15.4.1. Both the full model and the reduced model need to be specified using the lavaan Rosseel (2012) syntax as shown in the figure. Note that the population parameter values are also provided in the full model directly. The reduced model is similar to the full model with the paths to be tested removed. By clicking the "Calculate" button, both the effect

	RMSEA (ϵ)	Degrees of freedom (d)	Effect size (δ)
Small	0.05	1	0.0025
		2	0.005
		4	0.01
		8	0.02
		16	0.04
Medium	0.08	1	0.0064
		2	0.0128
		4	0.0256
		8	0.0512
		16	0.1024
Large	0.1	1	0.01
		2	0.02
		4	0.04
		8	0.08
		16	0.16

Table 15.4.1: Effect size for SEM

size and the degrees of freedom are shown in the output. For this example, the effect size in this example is 0.054.

15.4.2 Method 2: through empirical data and a model

The effect size can also be estimated from an empirical set of data. In this case, the full model is fitted to the data. The expected covariance matrix based on the full model is used as the population covariance matrix. Then, the reduced model will be fitted to the expected covariance matrix. Therefore, to use the method, one needs to provide a set of empirical data as well as the full and reduced models. The dataset can be uploaded to the server directly. The data file has to be in text format with each entry separated by white spaces. In addition, the first row should be the variable names that match those used in the full and reduced model. An example is given in Figure 15.4.2. The data used in this example is available at http://psychstat.org/semdata.

15.5 Technical Details

Technical details behind the power calculation used here can be found in Satorra & Saris (1985). We outline the basic idea in this book. Let **S** denote an unbiased sample covariance matrix and θ denote parameters in a SEM model. Let Σ be the covariance matrix defined by the model with parameters θ. From SEM theory, we know that the statistic

$$\hat{W} = (n-1)\left[\log|\Sigma(\hat{\theta})| + \text{tr}(S\Sigma(\hat{\theta})^{-1}) - \log|S| - p\right]$$

Effect Size Calculator for SEM
1. Effect size from population models

The model with population parameters (Full model)

```
y1 ~ 0.4*x
y2 ~ 0.5*x + 0.2*y1
y3 ~ 0.4*x
y4 ~ 0.4*y1 + 0.4*y2 + 0.4*y3
y1 ~~ 0.84*y1
y2 ~~ 0.61*y2
y3 ~~ 0.85*y3
y4 ~~ 0.27*y4
x ~~ 1*x
```

The restricted model fit to the population covariance matrix (Reduced m(

```
y1 ~ x
y2 ~ x
y3 ~ x
y4 ~ y1 + y2 + y3
y1 ~~ y1
y2 ~~ y2
y3 ~~ y3
y4 ~~ y4
x ~~ x
```

Calculate

Effect size output

The effect size = **0.054**
The degrees of freedom = **4**
RMSEA = **0.05788915**

Figure 15.4.1: Effect size calculation for SEM with predefined population model and parameter values

follows a chi-squared distribution with degrees of freedom d asymptotically. The purpose is to test the hypothesis that

$$H_0 : \theta = \theta_0$$

vs

$$H_1 : \theta = \theta_1.$$

Under H_0, we have $P(\chi_d^2 > c_\alpha) = \alpha$ where c_α is the critical value under the chi-squared distribution with degrees of freedom d. Under H_1, $\hat{W}$ follows asymptotically a non-central chi-squared distribution with the non-centrality parameter λ. The statistical power is defined as $Power = P(\hat{W} > c_\alpha | H_1)$. Satorra & Saris (1985) showed that λ can be approximated by

$$\lambda \approx (n-1)[\log|\hat{\Sigma}_R| + \text{tr}(\Sigma_F \hat{\Sigma}_R^{-1}) - \log|\Sigma_F| - p]$$

Effect Size Calculator for SEM
2. From empirical data analysis

Upload data file:

| Choose File | effect-sem.txt | | Calculate |

The model with population parameters (Full model)

```
y1 ~ x
y2 ~ x + y1
y3 ~ x
y4 ~ y1 + y2 + y3
y1 ~~ y1
y2 ~~ y2
y3 ~~ y3
y4 ~~ y4
x ~~ x
```

The restricted model fit to the population covariance matrix (Reduced mo

```
y1 ~ x
y2 ~ x
y3 ~ x
y4 ~ y1 + y2 + y3
y1 ~~ y1
y2 ~~ y2
y3 ~~ y3
y4 ~~ y4
x ~~ x
```

Effect size output

The effect size = **0.033**
The degrees of freedom = **4**
RMSEA = **0.04513109**

where Σ_F and Σ_R are defined under H_1 and H_0, respectively. With this, we can define an effect size independent of sample size as

$$\delta = \lambda/(n-1).$$

15.6 *Power Analysis Based on RMSEA Using the MacCallum et al. (1996) Method*

The primary software interface for SEM based on RMSEA (Browne & Cudeck, 1992) using the MacCallum et al. (1996) method is shown in Figure 15.6.1. Within the interface, a user can supply different parameter values and select different options for power analysis. We now explain each input.

http://psychstat.org/rmsea

- *Sample size* is the total number of participants. Multiple sample sizes can be provided in two ways to calculate power for each sample size. First, multiple sample sizes can be supplied and separated by white spaces, e.g., 100 150 200 will calculate power for the three sample sizes 100, 150 and 200. Second, a sequence of sample sizes can be generated using the method `s:e:i` with `s` denoting the starting sample size, `e` as the ending sample size, and `i` as the interval. Note that the values are separated by colon ":". For example, 100:150:10 will generate a sequence of sample sizes - 100 110 120 130 140 150. The default sample size is 100.

- The *degrees of freedom* of the chi-squared test. Multiple degrees of freedom can be provided as for the sample size.

- The *RMSEA for H0* specifies the RMSEA under the null hypothesis. Usually, it is `0` but can be larger than `0`.

- The *RMSEA for H1* specifies the RMSEA under the alternative hypothesis. This RMSEA can be a value from a previous SEM analysis.

- The *Significance level* for power calculation is needed but usually set at the default value 0.05. Multiple significance levels can be supplied by separating them using white spaces.

- The *Power* specifies the desired statistical power. Multiple power values can be supplied by separating them using white spaces.

- The use of RMSEA can test both close fit or not-close fit by setting *Type of analysis*.

- The *Power* specifies the desired statistical power. Multiple power values can be supplied by separating them using white spaces.

- In addition to the required input, one can also request the plot of a power curve if multiple sample sizes or effect sizes are provided.

- A note (less than 200 characters) can also be provided to provide basic information on the analysis for future reference for registered users.

Among the following input, *Sample size*, *Degrees of freedom*, *RMSEA for H0*, *RMSEA for H1*, *Significance level*, and *Power*, one and only one can be left blank. However, note that a solution is not guaranteed when calculating a quantity other than power.

15.7 *Examples*

Using the example in Satorra & Saris (1985), we show how to conduct the power analysis based on RMSEA. In the example, Y_1, Y_2, Y_3, Y_4, and

SEM based on RMSEA

Parameters (Help)

Sample size	100
Degrees of freedom	4
RMSEA for H0	0
RMSEA for H1	0.116
Significance level	0.05
Power	
Type of analysis	Close fit ⬍
Power curve	No power curve ⬍
Note	SEM based on RMSEA

Calculate

Output

```
Power for SEM based on RMSEA

    n df rmsea0 rmsea1  power alpha
  100  4      0  0.116 0.4208  0.05

URL: http://psychstat.org/rmsea
```

Figure 15.6.1: Power for SEM with known sample size and effect size in Example 15.7.1

X satisfy the following population model

$$Y_1 = \gamma_1 X + \zeta_1$$
$$Y_2 = \gamma_2 X + \beta_{21} Y_1 + \zeta_2 \qquad (15.7.1)$$
$$Y_3 = \gamma_3 X + \zeta_3$$
$$Y_4 = \beta_{41} Y_1 + \beta_{42} Y_2 + \beta_{43} Y_3 + \zeta_4$$

where $\zeta_i, i = 1, 2, 3, 4$ follows a normal distribution with mean 0 and unknown variance ψ_i and X has mean 0 and variance 1. The parameter values are

Parameter	γ_1	γ_2	γ_3	β_{41}	β_{42}	β_{43}	β_{44}	ψ_1	ψ_2	ψ_3	ψ_4
value	.4	.5	.4	.4	.4	.4	.2	.84	.61	.84	.27

Suppose a model is fitted without β_{21}. Then, one can investigate the power to detect this using the likelihood ratio test.

For the example above, the full model fits perfect and thus the RMSEA for H0 is 0. The RMSEA for H1 when $\beta_{21} = 0.2$ is 0.116 and

Example 15.7.1: Calculate power given sample size and RMSEA

the degrees of freedom for the test is 4.[3] A researcher is interested in the power given that she can collect data from 100 participants.

[3] The method for calculating RMSEA under H1 will be explained later.

The input and output for calculating power for this study are given in Figure 15.6.1. In the *Sample size* field, we input 100 and in the *Degrees of freedom* field, we input 4. RMSEA for H0 and H1 are 0 and 0.116 respectively. The default *Significance level* 0.05 is used. The field for *Power* is left blank because it will be calculated. By clicking the "**Calculate**" button, the statistical power is given in the output. For the current analysis, the power is 0.4208.

Similar to the Satorra & Saris (1985) method, WebPower can generate a power curve with given sample sizes. For example, the input and output in Figure 15.7.1 calculate power with a sample size from 100 to 600 with an interval of 100. Note that in the *Sample size* field, the input is 100:600:100. In the output, the power for each sample size from 100 to 600 with the interval 100 is listed. In the input, we also choose "*Show power curve*" for the *Power curve* field. In the output, the power curve is displayed at the bottom of the output as shown in Figure 15.7.2. The power curve can be used for interpolation. For example, to get a power 0.8, a sample size of about 250 is needed.

Example 15.7.2: Power curve

In practice, a power 0.8 or higher is often desired. Given the power, the sample size can also be calculated as shown in Figure 15.7.3. In this situation, the *Sample size* field is left blank while in the *Power* field, the value 0.8 is input. In the output, we can see a sample size 223 is needed to obtain a power 0.8.

Example 15.7.3: Calculate sample size given power and RMSEA

Suppose a researcher does not have enough information to determine the size of RMSEA for H1. In this case, the researcher can find the minimum RMSEA to achieve certain power given the sample size. Such an effect can help the researcher determine whether it is meaningful to conduct the study. For example, if a very large RMSEA is needed, it might be difficult to achieve a significant result. In Figure 15.7.4, we calculate the effect size by providing both sample size and power as well as other information. Note that the RMSEA for H1 has to be at least 0.1736 to have a power 0.8 with the sample size 100.

Example 15.7.4: Calculate RMSEA for H1 given power and sample size

15.8 *Using WebPower for Power Analysis*

The power calculation for SEM based on the RMSEA method proposed by MacCallum et al. (1996) is conducted using the R function `wp.sem.rmsea`. The detail of the function is:

```
wp.sem.rmsea(n = NULL, df = NULL, rmsea0 = NULL, rmsea1 = NULL,
    power = NULL, alpha = 0.05, type=c('close','notclose'))
```

n: sample size
df: degrees of freedom
rmsea0: RMSEA for H0, usually 0
rmsea1: RMSEA for H1
alpha: significance level
power: statistical power
type: close fit or non-close fit

SEM based on RMSEA

Figure 15.7.1: Input and output for obtaining power curve for SEM in Example 15.7.2

Parameters (Help)

Sample size	100:600:100
Degrees of freedom	4
RMSEA for H0	0
RMSEA for H1	0.116
Significance level	0.05
Power	
Type of analysis	Close fit ‡
Power curve	Show power curve ‡
Note	SEM based on RMSEA

Calculate

Output

```
Power for SEM based on RMSEA

    n df rmsea0 rmsea1  power alpha
  100  4      0  0.116 0.4208  0.05
  200  4      0  0.116 0.7495  0.05
  300  4      0  0.116 0.9136  0.05
  400  4      0  0.116 0.9746  0.05
  500  4      0  0.116 0.9934  0.05
  600  4      0  0.116 0.9984  0.05


URL: http://psychstat.org/rmsea
```

The R input and output for the examples discussed in the previous section are given below.

```
> ## calculate power given sample size and rmsea
> wp.sem.rmsea(n = 100, df = 4, rmsea0 = 0, rmsea1 = .116, power
    = NULL, alpha = 0.05)

Power analysis for SEM based on RMSEA

    n df rmsea0 rmsea1     power alpha
  100  4      0  0.116 0.4208173  0.05

NOTE: Power analysis for SEM based on RMSEA

WebPower URL: http://psychstat.org/rmsea

> ## power curve
```

Figure 15.7.2: Input and output for obtaining power curve for SEM in Example 15.7.2

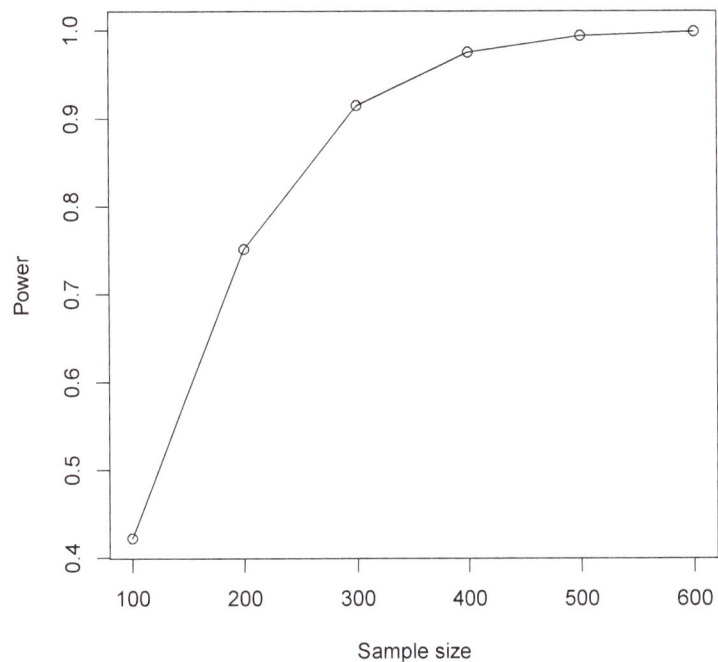

```
> res <- wp.sem.rmsea(n = seq(100,600,100), df = 4, rmsea0 = 0,
    rmsea1 = .116, power = NULL, alpha = 0.05)
> res

Power analysis for SEM based on RMSEA

    n df rmsea0 rmsea1     power alpha
  100  4      0  0.116 0.4208173  0.05
  200  4      0  0.116 0.7494932  0.05
  300  4      0  0.116 0.9135968  0.05
  400  4      0  0.116 0.9746240  0.05
  500  4      0  0.116 0.9933963  0.05
  600  4      0  0.116 0.9984373  0.05

NOTE: Power analysis for SEM based on RMSEA

WebPower URL: http://psychstat.org/rmsea

> plot(res)  ## generate power curve
>
> ## calculate the sample size given power
> wp.sem.rmsea(n = NULL, df = 4, rmsea0 = 0, rmsea1 = 0.116,
    power = 0.8, alpha = 0.05)
```

SEM based on RMSEA

Figure 15.7.3: Input and output for obtaining sample size for SEM in Example 15.7.3

Parameters (Help)

Sample size	
Degrees of freedom	4
RMSEA for H0	0
RMSEA for H1	0.116
Significance level	0.05
Power	0.8
Type of analysis	Close fit ⬍
Power curve	No power curve ⬍
Note	SEM based on RMSEA

Calculate

Output

```
   Power for SEM based on RMSEA

         n df rmsea0 rmsea1 power alpha
    222.7  4      0  0.116   0.8  0.05


URL: http://psychstat.org/rmsea
```

```
Power analysis for SEM based on RMSEA

          n df rmsea0 rmsea1 power alpha
   222.7465  4      0  0.116   0.8  0.05

NOTE: Power analysis for SEM based on RMSEA

WebPower URL: http://psychstat.org/rmsea

>
> ## calculate the rmsea1 given other information
> wp.sem.rmsea(n = 100, df = 4, rmsea0 = 0, rmsea1 = NULL, power
    = 0.8, alpha = 0.05)

Power analysis for SEM based on RMSEA

     n df rmsea0    rmsea1 power alpha
   100  4      0 0.1736082   0.8  0.05

NOTE: Power analysis for SEM based on RMSEA
```

SEM based on RMSEA

Parameters (Help)

Sample size	100
Degrees of freedom	4
RMSEA for H0	0
RMSEA for H1	
Significance level	0.05
Power	0.8
Type of analysis	Close fit
Power curve	No power curve
Note	SEM based on RMSEA

Calculate

Output

```
Power for SEM based on RMSEA

    n df rmsea0 rmsea1 power alpha
  100  4      0 0.1736   0.8  0.05


URL: http://psychstat.org/rmsea
```

```
WebPower URL: http://psychstat.org/rmsea
```

15.9 Effect Size

The effect size can be measured using RMSEA. According to Browne & Cudeck (1992), an RMSEA smaller than 0.05 indicates a close fit of the model. MacCallum et al. (1996) have used 0.01, 0.05, and 0.08 to indicate excellent, good, and mediocre fit, respectively. Large misfit indicates bigger effect in the power analysis sense. Therefore, we might consider 0.05 as small, 0.08 as medium, and 0.1 as large effect.

RMSEA can be calculated in a similar way as for the effect size of the Satorra & Saris (1985) method. First, from the full model, the model implied covariance matrix can be obtained as Σ_F. Then, the reduced model can be fitted to Σ_F. Suppose the estimated covariance matrix for the reduced model is $\hat{\Sigma}_R$. The effect size δ is for the Satorra & Saris (1985) method obtained as

$$\delta = \log|\hat{\Sigma}_R| + \mathrm{tr}(\Sigma_F\hat{\Sigma}_R^{-1}) - \log|\Sigma_F| - p$$

where p is the first dimension of Σ_F. The RMSEA ϵ is

$$\epsilon = \sqrt{\frac{\delta}{d}}$$

with d denoting the degrees of freedom. Note that RMSEA used here did not correct the sample size as in Browne & Cudeck (1992). This makes it independent of sample size but will be slightly different from the typically reported RMSEA although the difference should be small in general. WebPower effect size calculator for the Satorra & Saris (1985) method also outputs RMSEA.

15.10   Technical Details

Technical details behind the power calculation used here can be found in MacCallum et al. (1996). We only outline the basic idea. Let ϵ_0 and ϵ_1 be RMSEA under H_0 and H_1. Under H_0, the test statistic

$$\hat{W} = (n-1)\left[\log|\Sigma(\hat{\theta})| + \text{tr}(S\Sigma(\hat{\theta})^{-1}) - \log|S| - p\right]$$

follows a chi-squared distribution with degrees of freedom d and non-centrality parameter $\lambda_0 = nd\epsilon_0^2$. Note that $\lambda = 0$ if the model fits perfectly as the case in the Satorra & Saris (1985) method. Under H_1, the test statistic follows a chi-squared distribution with degrees of freedom d and non-centrality parameter $\lambda_1 = nd\epsilon_1^2$. Therefore, the statistical power for testing close fit is defined as

$$
\begin{aligned}
Power &= P(\hat{W} > c_\alpha|H_1) \\
&= 1 - [\chi^2_{d,\lambda_1}(c_\alpha)]^{-1} \\
&= 1 - \chi^2_{d,\lambda_1}[(\chi^2_{d,\lambda_0,\alpha})]^{-1}
\end{aligned}
$$

where $[\chi^2_{d,\lambda_0,\alpha}]^{-1}$ is the $100\alpha th$ percentile of the chi-squared distribution under H_0. For the not-close fit, the power is

$$
\begin{aligned}
Power &= P(\hat{W} < c_\alpha|H_1) \\
&= [\chi^2_{d,\lambda_1}(c_\alpha)]^{-1} \\
&= \chi^2_{d,\lambda_1}[(\chi^2_{d,\lambda_0,\alpha})]^{-1}.
\end{aligned}
$$

In the not-close fit situation, we expect that $\epsilon_1 < \epsilon_0$.

15.11   Exercises

1. A researcher wants to understand the effect of school climate and peer support on teacher burnout. The theoretical construct "burnout" can be estimated by three measures of emotional exhaustion (EE1, EE2, EE3). The construct of School Climate can be modeled by

three measures (CC2, CC3, CC4). The construct of Peer Support can be modeled by two measures (PS1 and PS2). School Climate has an influence on Peer Support and Burnout. Peer Support has an influence on Burnout. Burnout is the ultimate outcome of interest. Based on the literature, the researcher sketches the following path diagram.

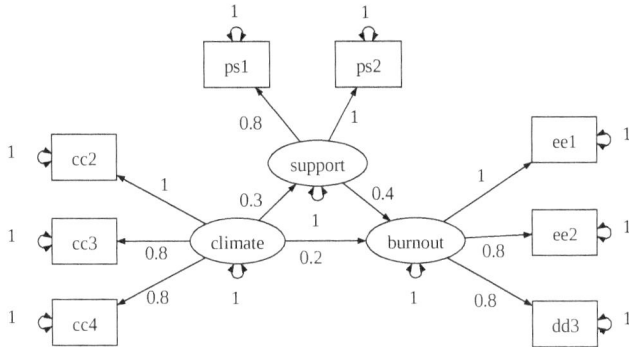

(a) Based on the path diagram, what is the effect size for the path from peer support to burnout?

(b) If the researcher can collect data from 200 participants, what is the power to detect the effect from peer support to burnout?

(c) Create a power curve for detecting the path from peer support to burnout. How many participants are needed to get a power 0.9?

Part II

Statistical Power Analysis based on Monte Carlo Simulation

16 Power Analysis for t-test with Non-normal Data and Unequal Variances

Han Du
Department of Psychology
University of California, Los Angeles

Zhiyong Zhang and Ke-Hai Yuan
Department of Psychology
University of Notre Dame

This chapter is based on Du et al. (2017b).

Power analysis is widely used for sample size determination (Cohen, 1988). With power analysis, an adequate but not "too large" sample size is determined to detect an existing effect. The conventional method for power analysis for the t-test is limited by two strict assumptions: normality and homogeneity (two-sample pooled-variance t-test). The two-sample separated-variance t-test (also known as the Welch's t-test; Welch, 1947) tolerates heterogeneity but still assumes normally distributed data. Thus, the corresponding exact power solution for the separated-variance t-test assumes normality with either numerical integration of noncentral density function or approximation (e.g., DiSantostefano & Muller, 1995; Moser et al., 1989).

Practical data in social, behavioral, and education research are rarely normal or homogeneous (Blanca et al., 2013; Cain et al., 2017; Micceri, 1989). This poses challenges on statistical power analysis for the t-test (Cain et al., 2017). To deal with the problems, we develop a general method to conduct power analysis for t-test through Monte Carlo simulation. The method can flexibly take into account non-normality in the one-sample t-test, two-sample t-test, and paired t-test, and unequal variances in two-sample t-test. We provide an R function as well as an online interface for implementing the proposed Monte Carlo based power analysis procedure.

One-sample t-test

The one-sample t-test concerns whether the population mean μ is different from a specific target value μ_0 (usually $\mu_0 = 0$). Thus the null hypothesis is

$$H_0 : \mu = \mu_0.$$

The alternative hypothesis can be either two-sided (H_{a1}) or one-sided (H_{a2} or H_{a3}):

$$H_{a1} : \mu \neq \mu_0,$$
$$H_{a2} : \mu > \mu_0, \text{ or}$$
$$H_{a3} : \mu < \mu_0.$$

The statistic given a sample size n,

$$t = \frac{\bar{y} - \mu_0}{s/\sqrt{n}},$$

follows a t distribution with degrees of freedom $n - 1$ under the normality assumption, where s is the sample standard deviation. When the normality assumption is violated, the t statistic does not follow a t distribution anymore. When sample size increases, the statistic approximately follows a normal distribution. However, power analysis is less meaningful with a huge sample size because the power would be close to 1.

Non-normality can take many forms. In this study, we focus on continuous variables with skewness and kurtosis different from those of a normal distribution (e.g., Cain et al., 2017). For non-normal data with an unknown distribution, it is extremely difficult to use an analytical formula to calculate power as in traditional power analysis. Instead, a Monte Carlo simulation method can be conveniently used (e.g., Muthén & Muthén, 2002; Zhang, 2014). The basic procedure of the Monte Carlo method is to first simulate the empirical null distribution of a chosen test statistic with the first four moments (Zhang, 2018) under the null distribution to get the empirical critical value for null hypothesis testing. Then the distribution of the test statistic under the alternative hypothesis can be simulated and the power can be estimated using the empirical distribution under the alternative hypothesis and the empirical critical value. The general method for SEM can be found in Yuan et al. (2017).

To use the Monte Carlo method, information regarding the first four moments is needed (Zhang, 2018). Specifically, we need the population mean (μ) and standard deviation (σ). In addition, we need the population skewness

$$\gamma_1 = E\left[\left(\frac{x - \mu}{\sigma}\right)^3\right] = \frac{\mu_3}{\sigma^3}$$

and kurtosis

$$\gamma_2 = E\left[\left(\frac{x-\mu}{\sigma}\right)^4\right] = \frac{\mu_4}{\sigma^4}.$$

For a normal distribution, the skewness is 0 and the kurtosis is 3. For testing the population mean, the means under the null and alternative hypotheses should be different, denoted by μ_0 and μ_1, respectively. However, we assume that the shapes of distributions under the null and alternative are the same with the same standard deviation, skewness, and kurtosis although they can be different. In practice, the population measures are unknown but they can be approximated based on meta-analysis or literature review (e.g., Schmidt & Hunter, 2014).

For the one-sample test, the following step-by-step procedure can be used to obtain the power for a given sample size n for testing

$$H_0 : \mu = \mu_0 \text{ vs. } H_1 : \mu = \mu_1.$$

1. Given the mean (μ_0), standard deviation (σ), skewness (γ_1), and kurtosis (γ_2), generate R_0 sets of non-normal data, each with the sample size n. R_0 should be sufficiently large and we recommend a minimum value 100,000.

2. Calculate the mean and variance for each of the R_0 datasets denoted as $\bar{y}_{0j}$ and $s^2_{0j}, j = 1, \ldots, R_0$. Calculate the statistics

    $$t^*_{0j} = \frac{\bar{y}_{0j} - \mu_0}{s_{0j}/\sqrt{n}}.$$

    Obtain the critical value c_α according to the pre-specified type I error rate α, typically 0.05, and the alternative hypothesis. For example, if the alternative hypothesis is H_{a2}, c_α is the $100(1-\alpha)$th percentile of t^*_{0j}.

3. Generate R_1 sets of non-normal data, each with the sample size (n), the mean (μ_1), standard deviation (σ), skewness (γ_1), and kurtosis (γ_2). We recommend a minimum value 1,000 for R_1.

4. Calculate the mean and variance for each dataset in Step (3) and denote them as $\bar{y}_{ai}$ and $s^2_{ai}, i = 1, \ldots, R_1$, and calculate the corresponding statistic
    $$t^*_{ai} = \frac{\bar{y}_{ai} - \mu_0}{s_{ai}/\sqrt{n}}.$$

5. The power is estimated as the proportion that t^*_{ai} is greater than the critical value c_α:
    $$\pi = \#(t^*_{ai} > c_\alpha)/R_1.$$

The Monte Carlo procedure works equally for the normal data, in which the data in Step (1) and (3) can be generated from normal distributions.

The procedure above also works for the paired samples where the population mean, standard deviation, skewness, and kurtosis of the difference scores are used.

16.2   *Two-sample t-test*

The two-sample t-test is used to test whether two independent population means are equal. The null hypothesis is

$$H_0 : \mu_1 = \mu_2.$$

The alternative hypothesis can be either two-sided (H_{a1}) or one-sided (H_{a2} or H_{a3}):

$$H_{a1} : \mu_1 \neq \mu_2,$$
$$H_{a2} : \mu_1 > \mu_2, \text{ or}$$
$$H_{a3} : \mu_1 < \mu_2.$$

The pooled-variance t statistic

$$t_{pooled} = \frac{\bar{y}_1 - \bar{y}_2}{\sqrt{\frac{(n_1-1)s_1^2+(n_2-1)s_2^2}{n_1+n_2-2}(1/n_1 + 1/n_2)}}$$

follows a t distribution with degrees of freedom $n_1 + n_2 - 2$, where n_1 and n_2 are sample sizes for the two independent samples. $\bar{y}_1$ and $\bar{y}_2$ are the sample means and s_1^2 and s_2^2 are the sample variances of the two groups, respectively. The pooled t-test assumes homogeneity and normality. When the variances of the two groups are not the same, the separated-variance t-test should be used where the test statistic

$$t = \frac{\bar{y}_1 - \bar{y}_2}{\sqrt{\frac{s_1^2}{n_1} + \frac{s_2^2}{n_2}}}$$

follows a t-distribution with the degrees of freedom

$$df = \frac{\left(\frac{s_1^2}{n_1} + \frac{s_2^2}{n_2}\right)^2}{\frac{(s_1^2/n_1)^2}{n_1-1} + \frac{(s_2^2/n_2)^2}{n_2-1}}.$$

When the normality assumption is violated, the distribution of the statistic is not a t distribution anymore. Therefore, the Monte Carlo based method could be used for power analysis.

As in one-sample t-test, we assume that the shapes of the population distribution for each group under the null and alternative hypotheses are the same with the same standard deviation, skewness, and kurtosis. The step-by-step procedure for the two-sample t-test power calculation with given sample sizes n_1 and n_2 for the two groups is given below.

1. Let μ_{10} and μ_{20} be the means of the two groups under the null hypothesis, typically, $\mu_{10} - \mu_{20} = 0$. Given the population means (μ_{10} and μ_{20}), standard deviations (σ_1 and σ_2), skewness values (γ_{11} and γ_{12}), and kurtosis values for two groups (γ_{21} and γ_{22}), generate R_0 sets of non-normal data, one with sample size n_1 and another with sample size n_2. We recommend a minimum value 100,000 for R_1.

2. For the R_0 sets of data from previously simulated data pool, calculate the mean and variance of each group for each dataset denoted as y_{01j}, y_{02j}, s_{01j}^2, and s_{02j}^2, $j = 1, \ldots, R_0$. Calculate the separated-variance test statistics

$$t_{0j}^* = \frac{\bar{y}_{01j} - \bar{y}_{02j}}{\sqrt{\frac{s_{01j}^2}{n_1} + \frac{s_{02j}^2}{n_2}}}.$$

   Obtain the critical value c_α according to the pre-specified type I error rate α and the alternative hypothesis.

3. Let μ_{11} and μ_{21} be the means of the two groups under the alternative hypothesis. Generate R_1 sets of non-normal data, each with the sample sizes (n_1 and n_2), means (μ_{11} and μ_{21}), standard deviations (σ_1 and σ_2), skewness values (γ_{11} and γ_{12}), and kurtosis values (γ_{21} and γ_{22}) for the two groups separately. We recommend a minimum value 1,000 for R_1.

4. Calculate the mean and variance of each group for each dataset denoted as $\bar{y}_{a1i}$, $\bar{y}_{a2i}$, s_{a1i}^2, and s_{a2i}^2, $i = 1, \ldots, R_1$. Calculate the separated-variance test statistics

$$t_{ai}^* = \frac{\bar{y}_{a1i} - \bar{y}_{a2i}}{\sqrt{\frac{s_{a1i}^2}{n_1} + \frac{s_{a2i}^2}{n_2}}}.$$

5. The power is estimated as the proportion that t_{ai}^* is greater than the critical value c_α:

$$\pi = \#(t_{ai}^* > c_\alpha)/R_1.$$

16.3   *Power Analysis Using R Package WebPower*

The Monte Carlo procedure for power analysis for the one-sample, paired sample, and two-sample analyses is implemented in an R package WebPower. Specifically, the function wp.mc.t is utilized.

The basic usage of the function wp.mc.t has the following form:

```
wp.mc.t(n = NULL, R0 = 1e+05, R1 = 1000, mu0 = 0, mu1 = 0,
    sd = 1, skewness = 0, kurtosis = 3, alpha = 0.05, type = c("
    two.sample",          "one.sample", "paired"), alternative =
    c("two.sided",          "less", "greater"))
```

In the function, n is the sample size; mu0, mu1, sd, skewness, and kurtosis are the mean under the null hypothesis, mean under the alternative hypothesis, standard deviation, skewness, and kurtosis, with the default values 0, 0, 1, 0, and 3, respectively. R0 and R1 specify the total number of replications under null and alternative hypotheses with the default values 100,000 and 1,000, respectively. alpha is the significance level with the default value 0.05. type specifies the type of analysis such as one-sample test or two-sample test, and alternative specifies the direction of the alternative hypothesis.

In a one-sample t-test, we are interested in whether the population mean is equal to 0 with a two-sided alternative hypothesis. The population distribution follows a *normal* distribution with the mean 0.5 and the standard deviation 1. Therefore, the default skewness 0 and kurtosis 3 are used here. To calculate the power with the sample size 20, the R input and output are given below. The power is 0.562 in this example.

Example 16.3.1: Calculate power for one sample t-test

```
> wp.mc.t(n=20 , mu0=0, mu1=0.5, sd=1, skewness=0, kurtosis=3,
    type = c("one.sample"), alternative = c("two.sided"))

One-sample t test power calculation

    n power mu0 mu1 sd skewness kurtosis alpha
   20 0.562   0 0.5  1        0        3  0.05

WebPower URL: http://psychstat.org/tnonnormal
```

In a paired t-test, we plan to test whether the matched pairs have equal means with the one-sided alternative hypothesis ($H_a : \mu_D > 0$). The mean, standard deviation, skewness, and kurtosis of the difference scores are 0.3, 1, 1, and 6 respectively. Therefore, the data are not normally distributed. To calculate statistical power with the sample size 40, the R input and output are given below. The obtained power is 0.664 in this example.

Example 16.3.2: Calculate power for paired t-test

```
> wp.mc.t(n=40 , mu0=0, mu1=0.3, sd=1, skewness=1, kurtosis=6,
    type = c("paired"), alternative = c("greater"))

Paired t test power calculation

    n power mu0 mu1 sd skewness kurtosis alpha
   40 0.664   0 0.3  1        1        6  0.05

NOTE: n is number of *pairs*

WebPower URL: http://psychstat.org/tnonnormal
```

Example 16.3.3: Calculate power for two-sample t test

In a two-sample independent t-test, we plan to examine whether two independent population means are equal with the one-sided alternative hypothesis ($H_a : \mu_1 - \mu_2 < 0$). The means for the two groups are 0.2 and 0.5, standard deviations for the two groups are 0.2 and 0.5, skewnesses for the two groups are 1 and 2, and kurtoses for the two groups are 4 and 6 respectively. To calculate the power with the sample size equal to 15 per group, the specification of the R function is as follows. The obtained power is 0.89 in this example.

```
> wp.mc.t(n=c(15, 15), mu1=c(0.2, 0.5), sd=c(0.2, 0.5), skewness=
    c(1, 2), kurtosis=c(4, 6), type = c("two.sample"),
    alternative = c("less"))

Two-sample t test power calculation

    n1 n2 power mean1 mean2 sd1 sd2 skewness1 skewness2 kurtosis1
        kurtosis2 alpha
    15 15  0.89   0.2   0.5 0.2 0.5         1         2         4
                4  0.05

NOTE: n is the sample size in *each* group

WebPower URL: http://psychstat.org/tnonnormal
```

16.4 Power Analysis Using WebPower Online Interface

An online interface can also be used to conduct the same power analysis as shown in Figure 16.4.1. The following information is needed to conduct the Monte Carlo power analysis.

http://psychstat.org/tnonnormal

- *Sample size* is the total number of participants. Multiple sample sizes can be provided in two ways to calculate power for each sample size. First, multiple sample sizes can be supplied and separated by white spaces, e.g., 100 150 200 will calculate power for the three sample sizes 100, 150 and 200. Second, a sequence of sample sizes can be generated using the method s:e:i with s denoting the starting sample size, e as the ending sample size, and i as the interval. Note the values are separated by colon ":". For example, 100:150:10 will generate a sequence of sample sizes - 100 110 120 130 140 150. The default sample size is 100.

- *Replication for H0 and H1* specifies the total number of replications under null and alternative hypotheses with the default value 100,000 and 1,000, respectively.

- *Mean for H0* provides the mean under the null hypothesis. For a two-sample test, it should have two values separated by a space.

- *Mean for H1* provides the mean under the alternative hypothesis. For a two-sample test, it should have two values separated by a space.

- *Standard deviation, Skewness,* and *Kurtosis* specify the corresponding statistics under the population. They are assumed to be the same under both null and alternative hypotheses but can be different.

- *Significance level* is set at the default value 0.05 but can be changed.

- *Type of test* can be a one-sample test or a two-sample test.

- *H1* specifies the direction of the alternative hypothesis.

- *Power curve* controls whether to generate a power curve if multiple sample sizes are used.

To use the interface, simply input the needed information and click on "Calculate". For example, using the information in Example 16.3.1, we get a power of 0.557. Note that since the power is estimated based on simulation, each time, one might get a slightly different value.

Testing mean with non-normal data

Parameters (Help)

Sample size	20
Replication for H0	100000
Replication for H1	1000
Mean for H0	0
Mean for H1	0.5
Standard deviation	1
Skewness	0
Kurtosis	3
Significance level	0.05
Type of test	One sample ▾
H1	Two sided ▾
Power curve	No power curve ▾
Note	Testing mean with non-norr

Calculate

Figure 16.4.1: Power analysis for t-test with non-normal data

One-sample t test power calculation

```
One-sample t test power calculation

    n power mu0 mu1 sd skewness kurtosis alpha
   20 0.557   0 0.5  1        0        3  0.05

URL: http://psychstat.org/tnonnormal
```

17 Statistical Power for SEM and Mediation with Non-normal Data based on Monte Carlo Simulation

Zhiyong Zhang
Department of Psychology
University of Notre Dame

Structural equation modeling and mediation analysis are widely used in the social and behavioral sciences. In previous chapters, we have discussed how to conduct power analysis for SEM and a simple mediation model based on analytical solutions. In this chapter, we focus on how to deal with non-normal data and for a general set of mediation models. Our discussion focuses on mediation models but SEM can be viewed as a mediation model without a mediation effect being investigated in this chapter. Therefore, the same procedure works for SEM.

This chapter uses examples from Zhang (2014).

17.1 Monte Carlo Based Statistical Power Analysis

In this section, we first present the Monte Carlo based method. For better illustration, we focus our discussion on a simple mediation model even though the method applies to more complex models, as shown in our examples. Figure 17.1.1 displays the path diagram of the simple mediation model. In the figure, x, m, and y represent the independent or input variable, the mediation variable, and the dependent or outcome variable, respectively. In this model, the total effect of x on y, c'+ a∗b, consists of the direct effect c' and the mediation effect θ = a∗b, the multiplication of the direct effect of x on m and the direct effect of m on y. The mediation effect is also called the indirect effect because it is the effect of x on y indirectly through m.

Statistical power analysis for mediation can be viewed as concerning

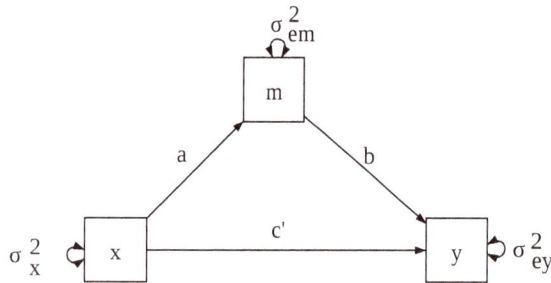

a test whether the mediation effect (θ) is significantly different from 0. More specifically, we have the null and alternative hypothesis

$$H_0 : \theta = \theta_0 \text{ vs. } H_1 : \theta = \theta_1,$$

where θ_0 is usually 0 and θ_1 represents a given effect size. By its definition, the statistical power (π) is

$$\pi = \Pr(\text{reject } H_0 | H_1). \quad (17.1.1)$$

In addition to the use of null hypothesis testing, the power can be calculated using the confidence intervals. This is based on the equivalence of confidence intervals and null hypothesis testing for testing a hypothesis (e.g., Hoenig & Heisey, 2001; Meehl, 1997). That is, if a 1-α confidence interval does not include the null hypothesis value, one can infer a statistically significant result at the significance level α (e.g., Daly, 1991). More specifically, let $[l, u]$ denote the confidence interval of the mediation effect θ. The power is then

$$\pi = \Pr(0 \notin [l, u] | H_1). \quad (17.1.2)$$

In practice, the power π can be difficult to calculate analytically especially for complex mediation models. However, it can be estimated using the relative frequency of rejecting the null hypothesis in Monte Carlo simulation following Algorithm 1. The algorithm has been widely applied in the literature of statistical power analysis for both mediation analysis and other analysis (e.g., Cheung, 2007; Fritz & MacKinnon, 2007; Fritz et al., 2012; Hayes & Scharkow, 2013; MacKinnon et al., 2004; Muthén & Muthén, 2002; Thoemmes et al., 2010; Zhang & Wang, 2009, 2013).

A critical component of such a Monte Carlo algorithm is the choice of the method for constructing the confidence interval of the mediation

---

Algorithm 1 Monte Carlo simulation algorithm for statistical power

1. Form a mediation model based on the hypothesized theory and set up the population parameters for the mediation model. The parameter values can be decided from previous studies in the literature or a pilot study.

2. Generate a data set with sample size n based on the model and its population parameter values.

3. Test the significance of a mediation effect by forming a confidence interval using the generated data.

4. Repeat Steps 2 and 3 for R times where R is the number of Monte Carlo replications.

5. Suppose among the R replications, the mediation effect is significant for r times. Then the power for detecting the mediation effect given the sample size n is r/R.

---

effect. Our R package allows three types of confidence intervals: the normal confidence interval, the robust confidence interval, and the bootstrap confidence interval although we recommend the use of the bootstrap confidence interval.

17.1.1 *Normal confidence interval*

In mediation analysis, model parameters and their covariance can be estimated using the maximum likelihood method. Under the normal data assumption, the estimated model parameters follow a multivariate normal distribution. For example, for the simple mediation model, $\hat{a}$ and $\hat{b}$, estimates of a and b, have a bivariate normal distribution with the covariance matrix $\begin{pmatrix} \hat{\sigma}_a^2 & \hat{\sigma}_{ab} \\ \hat{\sigma}_{ab} & \hat{\sigma}_b^2 \end{pmatrix}$ where $\hat{\sigma}_a^2$, $\hat{\sigma}_b^2$, and $\hat{\sigma}_{ab}$ are the estimated variances and covariance of $\hat{a}$ and $\hat{b}$. Using the delta method, $\hat{\theta} = \hat{a}\hat{b}$ is normally distributed with mean $\theta = ab$ and variance $\hat{b}^2\hat{\sigma}_a^2 + 2\hat{a}\hat{b}\hat{\sigma}_{ab} + \hat{a}^2\hat{\sigma}_b^2$ (p.298, Sobel, 1982). The $1 - \alpha$ confidence interval for ab can be constructed as

$$[\hat{a}\hat{b} + \Phi^{-1}(\alpha/2) \times \widehat{se}(\hat{a}\hat{b}), \hat{a}\hat{b} + \Phi^{-1}(1 - \alpha/2) \times \widehat{se}(\hat{a}\hat{b})], \quad (17.1.3)$$

where Φ is the standard normal cumulative distribution function and therefore $\Phi^{-1}(\alpha)$ gives the 100αth percentile of the standard normal distribution. For example, for the 95% confidence interval, $\Phi^{-1}(\alpha/2) = \Phi^{-1}(.05/2) = \Phi^{-1}(.025) \approx -1.96$ and $\Phi^{-1}(1 - \alpha/2) = \Phi^{-1}(.975) \approx 1.96$. $\widehat{se}(\hat{a}\hat{b}) = \sqrt{\hat{b}^2\hat{\sigma}_a^2 + 2\hat{a}\hat{b}\hat{\sigma}_{ab} + \hat{a}^2\hat{\sigma}_b^2}$ is the standard error of $\hat{a}\hat{b}$. We refer to this interval as the normal confidence interval. Note a power analysis based on the normal confidence interval is the same as the use of the Sobel test.

17.1.2 Robust confidence interval

When data are not normally distributed, the standard error estimates of the parameter estimates of the mediation models are not consistent. Therefore, the confidence interval in Equation (17.1.3) is problematic. However, if the fourth moments (or kurtosis) of the non-normal data still exist, the robust Sandwich-type standard errors are consistent and can be used (Zu & Yuan, 2010). Therefore, replacing the normal standard error with the Sandwich-type standard error in Equation (17.1.3), we obtain a robust confidence interval for the mediation effect.

17.1.3 Bootstrap confidence interval

Both the normal and robust confidence intervals are based on asymptotic theory and they might not perform well in finite sample experiments (e.g., MacKinnon et al., 2004; Zu & Yuan, 2010). In the literature, confidence intervals constructed using the bootstrap method have been shown to perform better under many studied conditions (e.g., Cheung, 2007; Fritz & MacKinnon, 2007; Fritz et al., 2012; Hayes & Scharkow, 2013; MacKinnon et al., 2004; Preacher & Hayes, 2004; Shrout & Bolger, 2002). Algorithm 2 can be followed to construct a bootstrap confidence interval.

Algorithm 2 Bootstrap confidence interval algorithm

1. Using the original data set (Sample size = n) as a population, draw a bootstrap sample of n persons randomly with replacement.

2. With the bootstrap sample, estimate model parameters and compute estimated mediation effects.

3. Repeat Steps 1 and 2 for a total of B times. B is the number of bootstrap samples.

4. The bootstrap confidence intervals of model parameters and mediation effects are constructed.

Different bootstrap confidence intervals have been used for the bootstrap method in the literature of mediation analysis (e.g., Cheung, 2007; Fritz & MacKinnon, 2007; Fritz et al., 2012; Hayes & Scharkow, 2013; MacKinnon et al., 2004). Let θ denote a population mediation effect, $\hat{\theta}$ denote the estimate of θ from the original data, and

$\hat{\theta}^b, b = 1, \ldots, B$ denote its estimate for the bth bootstrap sample. A $100(1-\alpha)\%$ bootstrap confidence interval is formed in the following ways. First, the percentile bootstrap confidence interval can be constructed by $[\hat{\theta}^b(\alpha/2), \hat{\theta}^b(1-\alpha/2)]$ for a parameter with $\hat{\theta}^b(\alpha)$ denoting the 100αth percentile of the B bootstrap estimates. Second, the bias-corrected bootstrap confidence interval can be constructed as $[\hat{\theta}^b(\tilde{\alpha}_l), \hat{\theta}^b(\tilde{\alpha}_u)]$ where $\tilde{\alpha}_l$ and $\tilde{\alpha}_u$ are used to get the quantiles and are calculated by

$$\tilde{\alpha}_l = \Phi[2z_0 + \Phi^{-1}(\alpha/2)] \tag{17.1.4}$$

and

$$\tilde{\alpha}_u = \Phi[2z_0 + \Phi^{-1}(1-\alpha/2)] \tag{17.1.5}$$

with

$$z_0 = \Phi^{-1}\left[\frac{\text{number of times that } \hat{\theta}^b < \hat{\theta}}{B}\right]. \tag{17.1.6}$$

17.2   R Package

The proposed method in the above section is implemented in the free, open-source R package WebPower. The package can conduct power analysis based on the normal, robust, and bootstrap confidence intervals. We now illustrate the use of the package through a simple mediation model shown in Figure 17.2.1. The values in the figure are population parameters that can be decided from a pilot study or the existing literature. In this example, we choose the parameter values to represent a medium mediation effect. Some of the values are labeled using a, b, and cp. For demonstration, suppose we are interested in the power of the mediation effect ab=a*b and the total effect abc=a*b+cp.

Figure 17.2.1: An example mediation model with population parameters

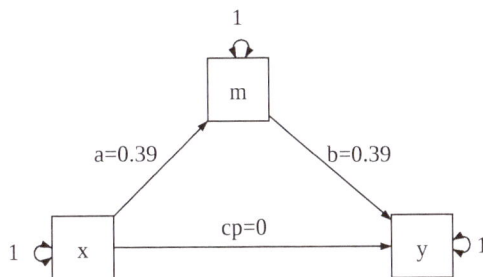

There are two functions in WebPower for power analysis with non-normal data for SEM and mediation: wp.mc.sem.basic and wp.mc.sem.boot.

The first function wp.mc.sem.basic estimates power based on the Sobel test either with regular standard error estimates or the robust standard error estimates. Since it does not require bootstrap, it runs much faster than the second function which is based on the bootstrap method for testing the significance of the mediation effects and model parameters.

To use the functions, one needs to specify the mediation model and the mediation effect. The package uses the lavaan (Rosseel, 2012) model specification method but with some specific requirements. For example, for the simple mediation model, it is specified as below.

```
demo = "
    y ~ cp*x + start(0)*x + b*m + start(.39)*m
    m ~ a*x+start(.39)*x
    x ~~ start(1)*x
    m ~~ start(1)*m
    y ~~ start(1)*y
"
```

First, the name of the model is demo in R. Everything about the model is given in a pair of quotation marks. Each path in the model is described using a line of statement. For example, m ~ a*x + start(.39)*x means that m regresses on x with the coefficient 0.39 as in start(0.39). Because the coefficient has a label a, it is also specified in the equation. The statement x ~~ start(1)*x means that the variance for x is 1. More generally, the regression relationships are specified using ~ and variance and covariance are specified using ~~. More about model specification can be found in Rosseel (2012).

For SEM, only the model is needed. But for mediation analysis, we also need to tell the package the mediation effects under evaluation. In this example, the mediation effect ab and the total effect abc are of interest to us. They can be specified as

```
mediation = "
    ab := a*b
    abc:= a*b + cp
"
```

The notation := means to calculate the indirect effect ab as the product of parameter a and b, where the labels on the right hand of ":=" should be consistent with those used in the model statement demo. Similarly, the total effect is calculated.

Only the labels for the parameters that will appear in the calculation of the mediation effect are necessary to use in the model specification part. For example, for the variance parameters, no labels are used. By default, the variance parameters will be set at 1. Therefore, in this example, the specifications of the three variance parameters are not required.

To conduct power analysis based on Sobel test, the following R code can be used. Note that the only difference between the regular and robust method is the use of the argument se="robust".

```
sobel.regular = wp.mc.sem.basic(model=demo, indirect=mediation,
    nobs=100, nrep=1000, parallel="snow", skewness=c(0, 0, 1.3),
    kurtosis=c(0, 0, 10), ovnames=c("x", "m", "y"))

sobel.robust = wp.mc.sem.basic(model=demo, indirect=mediation,
    nobs=100, nrep=1000, se="robust", parallel="snow", skewness=c
    (0, 0, 1.3), kurtosis=c(0, 0, 10), ovnames=c("x", "m", "y"))
```

To use the bootstrap method, the following R code is used.

```
mediation.boot = wp.mc.sem.boot(model=demo, indirect=mediation,
    nobs=100, nrep=1000, nboot=2000, parallel="snow", skewness=c
    (0, 0, 1.3), kurtosis=c(0, 0, 10), ovnames=c("x", "m", "y"))
```

The two functions have many arguments.

- model tells the mediation or SEM model to be used. It is required.

- indirect specifies the mediation effects of interest. It is not required.

- nobs is the sample size used. By default, the power is calculated for a sample size of 100.

- nrep specifies the number of Monte Carlo simulation replications in the calculation of power with a default 1000.

- nboot is the number of bootstrap with a default 1000 if the bootstrap method is used.

- parallel allows parallel computing. By setting parallel='snow', it uses the R package snowfall for automatic parallelization. By default, all cores available on a computer are used to speed up calculation.

- If one suspects the data will be non-normal, the skewness and kurtosis for the observed variables can be provided using skewness and kurtosis.

- When specifying non-normal data, the observed variable names (ovnames) should also be provided to match the order of the skewness and kurtosis statistics.

The results of the power analysis for both functions are similar and can be summarized into a table using the function summary(mediation.boot). The results include the following information for each parameter and mediation effect in the model.

- The column True lists the population parameter values.

- The column Estimate presents the average parameter estimates across all replications.

- The column MSE is the average bootstrap standard error.

- The column SD is the standard deviation of the parameter estimates across all replications.

- The column Power gives the power to detect whether a parameter is significant.

- The column Power.se provides the standard error of the estimated statistical power. The power for the mediation effect is listed at the end of the table entitled "Indirect/Mediation effects".

- The column Coverage presents the empirical coverage probability of the bias-corrected bootstrap confidence interval.

For the current example, the power to detect the mediation effect with a sample size 100 is about 0.935 using the percentile bootstrap confidence interval . If a researcher targets a power of 0.8, he/she can reduce the current sample size for another power calculation.

```
                     True  Estimate   MSE     SD     Power Power.se Coverage
Regressions:
  math ~
    ME      (cp)    0.000   0.001    0.107   0.111   0.070   0.008   0.930
    HE      (b)     0.390   0.395    0.109   0.119   0.971   0.005   0.928
  HE ~
    ME      (a)     0.390   0.394    0.101   0.103   0.966   0.006   0.938

Variances:
    math            1.000   0.975    0.262   0.351   1.000   0.000   0.791
    HE              1.000   0.982    0.135   0.140   1.000   0.000   0.915

Indirect/Mediation effects:
    ab              0.152   0.156    0.061   0.064   0.935   0.008   0.933
    abc             0.152   0.158    0.108   0.111   0.346   0.015   0.927
```

17.3   Examples

In this example, the model with its population parameter values in Figure 17.2.1 is used to explore whether the relationship between mothers' education (ME) and children's mathematical achievement (math) is mediated by home environment (HE; Zhang & Wang, 2013). In generating the non-normal data, the skewness is set at -0.3, -0.7, and 1.3, and the kurtosis is set at 1.5, 0, and 5 for ME, HE, and math, respectively. The skewness and kurtosis statistics are determined according to real data used in Zhang & Wang (2013). The power for the sample size 100 is estimated. The focus is the mediation effect ab. The R input and output for the power analysis are given below. From the output, we can see that the power for the mediation effect is 0.963.

Example 17.3.1: Simple Mediation Analysis

```
> ex1model<-'
+ math ~ c*ME+start(0)*ME + b*HE+start(.39)*HE
+ HE ~ a*ME+start(.39)*ME
+ '
```

```
>
> indirect<-'ab:=a*b'
>
> boot.non.normal<-wp.mc.sem.boot(ex1model, indirect, 100, nrep=2000, nboot=2000,
      parallel='snow', skewness=c(-.3, -.7, 1.3), kurtosis=c(1.5, 0, 5), ovnames=c('
      ME', 'HE', 'math'), ncore=60)
[1] 0.0665 0.9750 0.9695 0.9970 1.0000 1.0000 0.9630
> summary(boot.non.normal)
Basic information:

  Esimation method                               ML
  Standard error                           standard
  Number of requested bootstrap                2000
  Number of requested replications             2000
  Number of successful replications            2000

                  True   Estimate    MSE      SD    Power  Coverage
Regressions:
  math ~
    ME      (c)   0.000   0.000    0.108    0.112   0.067   0.933
    HE      (b)   0.390   0.389    0.101    0.103   0.975   0.928
  HE ~
    ME      (a)   0.390   0.390    0.101    0.103   0.970   0.930

Variances:
    math          1.000   0.974    0.213    0.260   0.997   0.869
    HE            1.000   0.980    0.136    0.143   1.000   0.922

Indirect/Mediation effects:
    ab            0.152   0.152    0.057    0.058   0.963   0.942
```

A power curve is useful to graphically display how power changes with sample size (e.g., Zhang & Wang, 2009). Using the model shown in Figure 17.3.1, we show how to generate a power curve. The substantive idea of the model in Figure 17.3.1 is that the relationship between age and eduction and the performance on the everyday problem solving test (ept) is mediated by memory ability measured by the Hopkins Verbal Learning Test (hvltt) and reasoning ability measured by three reasoning tests including word series (ws), letter sets (lt), and letter series (ls) tests (see Zhang & Wang, 2013). The population model parameter values are also displayed in the figure. The R input and output for the analysis are given below.

Example 17.3.2: Mediation analysis with a latent mediator (Power curve)

```
> ex2model<-'
+ ept ~ start(.4)*hvltt + b*hvltt + start(0)*age + start(0)*edu
    + start(2)*R
+ hvltt ~ start(-.35)*age + a*age + c*edu + start(.5)*edu
+ R ~ start(-.06)*age + start(.2)*edu
+ R =~ 1*ws + start(.8)*ls + start(.5)*lt
+ age ~~ start(30)*age
+ edu ~~ start(8)*edu
+ age ~~ start(-2.8)*edu
+ hvltt ~~ start(23)*hvltt
+ R ~~ start(14)*R
+ ws ~~ start(3)*ws
+ ls ~~ start(3)*ls
+ lt ~~ start(3)*lt
+ ept ~~ start(3)*ept
+ '
```

```
>
> indirect<-'ind1 := a*b + c*b'
>
> nobs <- seq(100, 2000, by=200)
>
> res <- wp.mc.sem.power.curve(model=ex2model, indirect=indirect,
nobs=nobs, type='boot', parallel='snow', ncore=60)
```

The power curve displays the power in detecting the effect of age and education on ept that is mediated by hvltt (a*b+c*b) for sample size from 100 to 1900 with an interval of 200. The plot shows that to get a power 0.8, a sample size about 1,500 is needed. Note that a power curve can be used to obtain power for a given sample size through interpolation, although the results might not be as accurate.

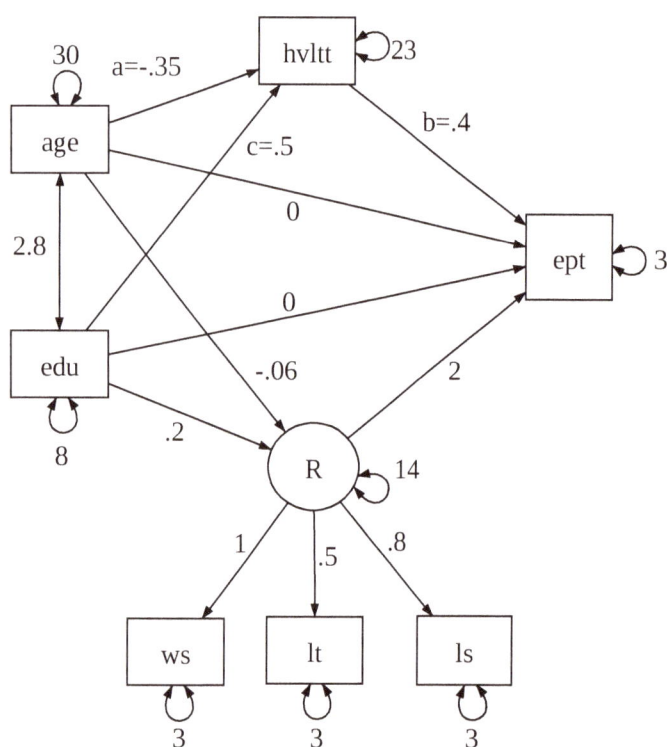

Figure 17.3.1: A multiple-mediator mediation model with population parameter values used in Example 17.3.2.

Thoemmes et al. (2010) considered a multiple group mediation model shown in Figure 17.3.3. Different from the simple mediation model in Figure 17.2.1, the mediator m is measured as a latent variable by three observed variables, m1, m2, and m3. Furthermore, two groups are considered with varying mediation effects. Specifically, the mediation effect for the first group is a1*b1 = 0.26 and for the second group is a2*b2 = 0.10. This implies a moderated mediation because the mediation effects are different for the two groups. The moderated

Example 17.3.3: Calculate power for multiple group mediation analysis

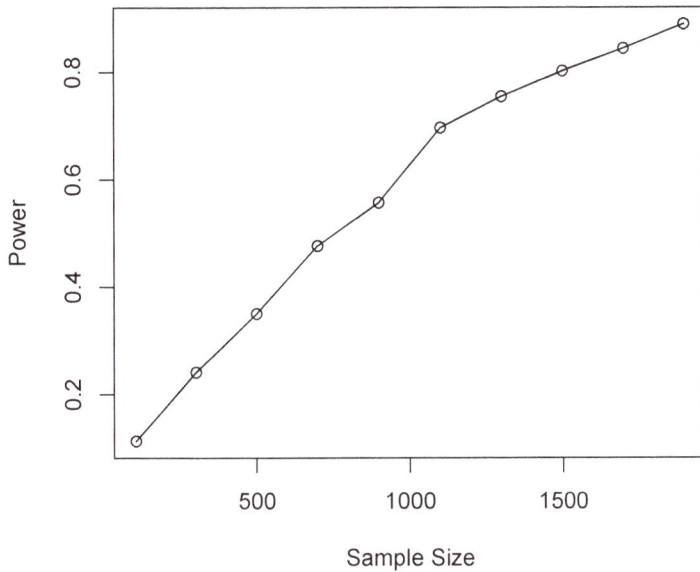

mediation can be evaluated using a1*b1 - a2*b2. The sample size for the first group is 400 and for the second group 200.

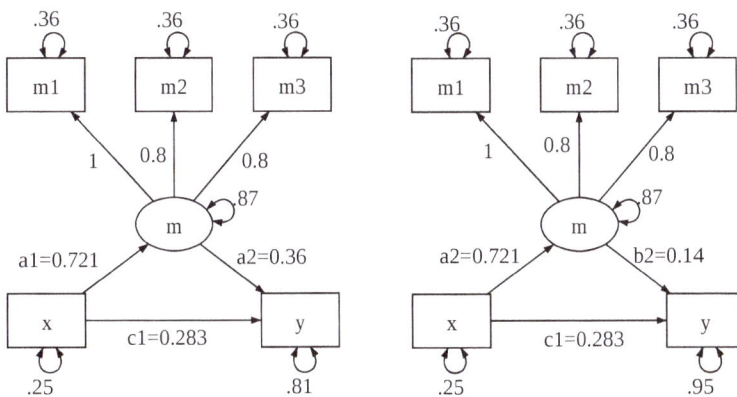

Suppose in this example, we are interested in the power to detect the mediation effect med1 = a1*b1 and med2 = a2*b2 as well as their difference diffmed = a1*b1 - a2*b2. The R input and output for the analysis are given below. Clearly, the power is 1 for med1, 0.444 for med2, and 0.473 for diffmed.

```
> ex3model<-"
+ y ~ start(c(.283, .283))*x + c(c1,c2)*x + start(c(.36, .14))*m +c(b1,b2)*m
+ m ~ start(c(.721, .721))*x + c(a1,a2)*x
+ m =~ c(1,1)*m1 + start(c(.8,.8))*m2 + start(c(.8,.8))*m3
+ x ~~ start(c(.25, .25))*x
```

```
+ y ~~ start(c(.81, .95))*y
+ m ~~ start(c(.87, .87))*m
+ m1 ~~ start(c(.36, .36))*m1
+ m2 ~~ start(c(.36, .36))*m2
+ m3 ~~ start(c(.36, .36))*m3
+ "
>
> indirect<-'
+ med1 := a1*b1
+ med2 := a2*b2
+ diffmed := a1*b1 - a2*b2
+ '
>
> bootstrap<-wp.mc.sem.boot(ex3model, indirect, nobs=c(400,200), nrep=2000,
nboot=1000, parallel='snow', ncore=60)

> summary(bootstrap)
Basic information:

  Esimation method                        ML
  Standard error                    standard
  Number of requested bootstrap         1000
  Number of requested replications      2000
  Number of successful replications     2000
```

Group 1 [1]:

		True	Estimate	MSE	SD	Power	Coverage
Latent variables:							
m =~							
m1		1.000	1.000	0.000	0.000	1.000	0.000
m2		0.800	0.799	0.047	0.048	1.000	0.938
m3		0.800	0.801	0.047	0.048	1.000	0.944
Regressions:							
y ~							
x	(c1)	0.283	0.282	0.099	0.100	0.810	0.943
m	(b1)	0.360	0.361	0.055	0.055	1.000	0.949
m ~							
x	(a1)	0.721	0.720	0.104	0.100	1.000	0.958
Intercepts:							
m1		0.000	0.000	0.055	0.056	0.054	0.946
m2		0.000	0.000	0.048	0.049	0.053	0.947
m3		0.000	0.001	0.048	0.050	0.061	0.939
y		0.000	0.000	0.048	0.048	0.051	0.949
x		0.000	-0.000	0.025	0.025	0.056	0.945
m		0.000	0.000	0.000	0.000	0.000	0.000
Variances:							
x		0.250	0.250	0.018	0.017	1.000	0.951
y		0.810	0.803	0.058	0.059	1.000	0.946
m		0.870	0.869	0.088	0.092	1.000	0.930
m1		0.360	0.358	0.047	0.046	1.000	0.947
m2		0.360	0.359	0.036	0.036	1.000	0.946
m3		0.360	0.358	0.036	0.035	1.000	0.955

Group 2 [2]:

		True	Estimate	MSE	SD	Power	Coverage
Latent variables:							
m =~							
m1		1.000	1.000	0.000	0.000	1.000	0.000
m2		0.800	0.801	0.069	0.070	1.000	0.940
m3		0.800	0.802	0.069	0.068	1.000	0.944
Regressions:							
y ~							
x	(c2)	0.283	0.285	0.151	0.154	0.469	0.941
m	(b2)	0.140	0.138	0.083	0.084	0.393	0.943
m ~							
x	(a2)	0.721	0.724	0.148	0.147	0.998	0.938

Intercepts:

m1	0.000	-0.000	0.078	0.079	0.052	0.948
m2	0.000	0.001	0.068	0.067	0.051	0.949
m3	0.000	0.001	0.068	0.067	0.052	0.948
y	0.000	0.001	0.070	0.070	0.051	0.949
x	0.000	0.001	0.035	0.036	0.057	0.943
m	0.000	0.000	0.000	0.000	0.000	0.000

Variances:

x	0.250	0.248	0.024	0.024	1.000	0.945
y	0.950	0.937	0.093	0.094	1.000	0.946
m	0.870	0.863	0.125	0.125	1.000	0.944
m1	0.360	0.355	0.068	0.068	1.000	0.939
m2	0.360	0.356	0.051	0.052	1.000	0.934
m3	0.360	0.354	0.051	0.052	1.000	0.933

Indirect/Mediation effects:

med1	0.260	0.260	0.054	0.052	1.000	0.952
med2	0.101	0.100	0.065	0.065	0.444	0.939
diffmed	0.159	0.160	0.085	0.085	0.473	0.947

Example 17.3.4: Calculate power for a longitudinal mediation model

Maxwell & Cole (2007) have recommended the use of longitudinal mediation models in mediation analysis because of the involvement of causal process in mediation. Figure 17.3.4 is a longitudinal mediation model derived from Figure 3 of Maxwell & Cole (2007) with population parameter values calculated from Table 2 of Maxwell and Cole. In this example, each variable in the mediation model is measured three times repeatedly. The idea of longitudinal mediation is that the input variable at time 1 influences the mediator at time 2 which in turn affects the outcome variable at time 3. The mediation effect is then measured by a*b as in the cross-sectional mediation models. The power is 0.892 when the bootstrap method is utilized for a sample size of 50.

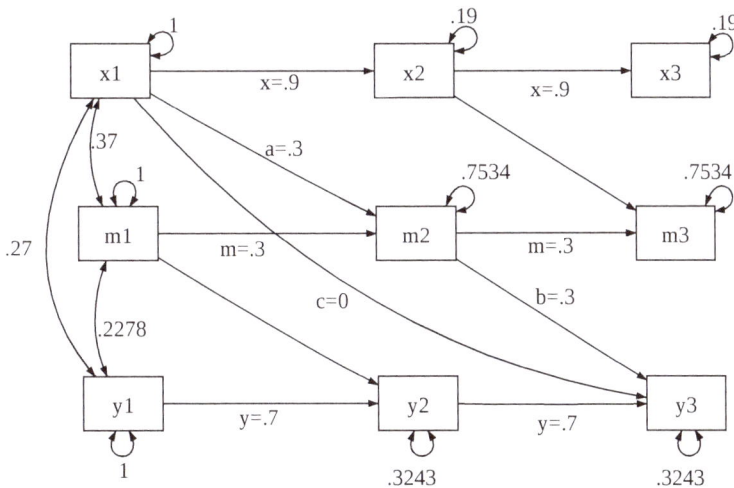

Figure 17.3.4: The path diagram for a longitudinal mediation model with population parameter values.

The R input and output for the analysis are given below.

```
> ex4model<-'
+ x2 ~ start(.9)*x1 + x*x1
+ x3 ~ start(.9)*x2 + x*x2
+ m2 ~ start(.3)*x1 + a*x1 + start(.3)*m1 + m*m1
+ m3 ~ start(.3)*x2 + a*x2 + start(.3)*m2 + m*m2
```

```
+ y2 ~ start(.3)*m1 + b*m1 + start(.7)*y1 + y*y1
+ y3 ~ start(.3)*m2 + b*m2 + start(.7)*y2 + y*y2 + start(0)*x1 + c*x1
+ x1 ~~ start(.37)*m1
+ x1 ~~ start(.27)*y1
+ y1 ~~ start(.2278)*m1
+ x2 ~~ start(.19)*x2
+ x3 ~~ start(.19)*x3
+ m2 ~~ start(.7534)*m2
+ m3 ~~ start(.7534)*m3
+ y2 ~~ start(.3243)*y2
+ y3 ~~ start(.3243)*y3
+ '
>
> indirect<-'ab:=a*b'
>
> bootstrap<-wp.mc.sem.boot(ex4model, indirect, nobs=50, nrep=1000, nboot=1000,
     parallel='snow', ncore=60)

> summary(bootstrap)
Basic information:

  Esimation method                            ML
  Standard error                        standard
  Number of requested bootstrap             1000
  Number of requested replications          1000
  Number of successful replications         1000
```

		True	Estimate	MSE	SD	Power	Coverage
Regressions:							
x2 ~							
x1	(x)	0.900	0.898	0.045	0.045	1.000	0.934
x3 ~							
x2	(x)	0.900	0.898	0.045	0.045	1.000	0.934
m2 ~							
x1	(a)	0.300	0.300	0.098	0.096	0.868	0.953
m1	(m)	0.300	0.293	0.096	0.099	0.843	0.932
m3 ~							
x2	(a)	0.300	0.300	0.098	0.096	0.868	0.953
m2	(m)	0.300	0.293	0.096	0.099	0.843	0.932
y2 ~							
m1	(b)	0.300	0.301	0.064	0.062	0.994	0.946
y1	(y)	0.700	0.694	0.063	0.064	1.000	0.935
y3 ~							
m2	(b)	0.300	0.301	0.064	0.062	0.994	0.946
y2	(y)	0.700	0.694	0.063	0.064	1.000	0.935
x1	(c)	0.000	-0.001	0.089	0.090	0.062	0.938
Covariances:							
x1 ~~							
m1		0.370	0.369	0.142	0.150	0.777	0.911
y1		0.270	0.261	0.138	0.145	0.467	0.920
m1 ~~							
y1		0.228	0.220	0.137	0.147	0.343	0.897
x3 ~~							
m3		1.000	-0.001	0.051	0.054	0.049	0.000
y3		1.000	0.001	0.034	0.036	0.054	0.000
m3 ~~							
y3		1.000	-0.002	0.066	0.069	0.058	0.000
Variances:							
x2		0.190	0.184	0.036	0.035	1.000	0.933
x3		0.190	0.187	0.036	0.039	1.000	0.919
m2		0.753	0.717	0.142	0.152	1.000	0.903
m3		0.753	0.718	0.140	0.144	1.000	0.926
y2		0.324	0.311	0.061	0.064	1.000	0.911
y3		0.324	0.306	0.060	0.065	1.000	0.890
x1		1.000	0.979	0.185	0.190	1.000	0.928
m1		1.000	0.984	0.188	0.201	1.000	0.911
y1		1.000	0.980	0.187	0.194	1.000	0.923
Indirect/Mediation effects:							
ab		0.090	0.090	0.036	0.035	0.892	0.953

18 | Statistical Power Analysis for Latent Change Score Models through Monte Carlo Simulation

Zhiyong Zhang
Department of Psychology
University of Notre Dame

Haiyan Liu
Psychological Sciences
University of California, Merced

Longitudinal data collection and data analysis are becoming a norm for psychological research (e.g., Grimm et al., 2016; McArdle & Nesselroade, 2014). Proposed by McArdle and colleagues, latent change score models (LCSMs) combine difference equations with growth curves to study change in longitudinal studies (e.g., McArdle, 2000; McArdle & Hamagami, 2001; Hamagami & McArdle, 2007; Hamagami et al., 2010). In such models, change is directly modeled, which is often the focus of a longitudinal study. In addition to the univariate LCSMs, bivariate LCSMs have also been proposed to model the inter-relationship between two growth processes (e.g., McArdle & Hamagami, 2001). Zhang & Liu (2018) proposed a Monte Carlo based method to determine the required sample size and/or the number of measurement occasions for both univariate and bivariate LCSMs. This method can obtain the power for testing each individual parameter of the models including the change rate and the coupling parameters. In this chapter, we illustrate how to conduct statistical power analysis for LCSMs in WebPower.

This chapter is based on Zhang & Liu (2018).

18.1 | A Univariate Latent Change Score Model

Let $Y[t]_n$ denote the data from the nth ($n = 1, \ldots, N$) participant at time t ($t = 1, \ldots, T$) of a sample consisting of N participants measured for

T times. The first part of a LCSM is the measurement error model in which an observed score $Y[t]_n$ is the sum of the latent true score $y[t]_n$ and the measurement error/uniqueness score $ey[t]_n$:

$$Y[t]_n = y[t]_n + ey[t]_n.$$

It is generally assumed that the error follows a normal distribution with mean 0 and variance $varey$. The second part of the model builds the relationship between consecutive latent true scores so that the current score at time t is equal to the sum of the true score at the previous time $t-1$ and the change, $dy[t]_n$, from time $t-1$ to time t:

$$y[t]_n = y[t-1]_n + dy[t]_n.$$

This effectively defines the change score as

$$dy[t]_n = y[t]_n - y[t-1]_n.$$

Note that in the classic LCSM, the relationship between consecutive latent true scores is deterministic although it is not required to be so. The third part of the model concerns the modeling of the difference scores. One way is to model the difference score at time t as the sum of a linear constant effect ys and the proportional change from time $t-1$ such that

$$dy[t]_n = ys_n + \beta_y \times y[t-1]_n,$$

where β_y is a compound rate of change.

The initial latent score and the linear constant change can be correlated. In the model, they are assumed to have a bivariate normal distribution

$$\begin{pmatrix} y0_n \\ ys_n \end{pmatrix} \sim MN \left[\begin{pmatrix} my0 \\ mys \end{pmatrix}, \begin{pmatrix} vary0 & vary0ys \\ vary0ys & varys \end{pmatrix} \right]$$

with MN denoting a multivariate, here bivariate, normal distribution. Therefore, the initial latent score follows a normal distribution with mean $my0$ and variance $vary0$ and the constant change also follows a normal distribution with mean mys and variance $varys$. The covariance between them is $vary0ys$. Using a path diagram, this model is portrayed in Figure 18.1.1.

18.2   A Bivariate Latent Change Score Model

A bivariate LCSM is first a combination of two univariate LCSMs. Above and beyond that, it allows the two processes represented by the LCSMs to interact with each other. Let $Y[t]_n$ and $X[t]_n$ denote the observed data on two variables, respectively, from the nth ($n =$

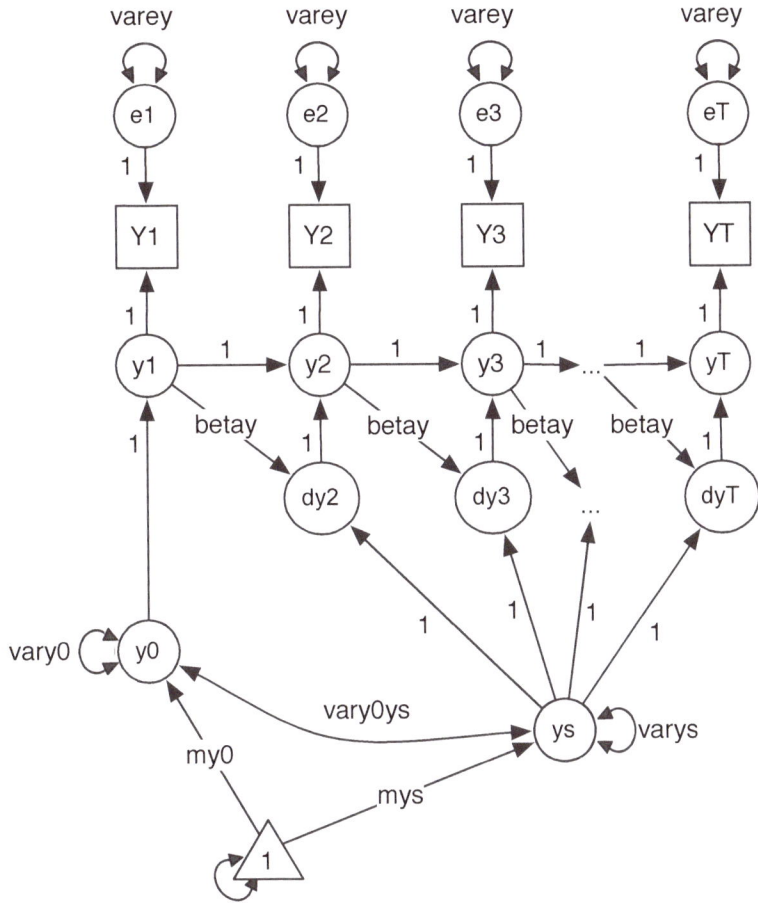

Figure 18.1.1: Path diagram for a univariate latent change score models

$1, \ldots, N$) participant at time t ($t = 1, \ldots, T$) of a sample consisting of N participants measured for T times. For the measurement error part of the model, we have

$$Y[t]_n = y[t]_n + ey[t]_n$$
$$X[t]_n = x[t]_n + ex[t]_n,$$

where $ey[t]_n$ follows a normal distribution with mean 0 and variance $varey$ and $ex[t]_n$ follows a normal distribution with mean 0 and variance $varex$. For the latent score from time $t-1$ to time t, we have

$$y[t]_n = y[t-1]_n + dy[t]_n$$
$$x[t]_n = x[t-1]_n + dx[t]_n,$$

with $dy[t]_n$ and $dx[t]_n$ denoting the latent change score for the two variables respectively.

The innovative part of the bivariate LCSM is to allow the latent score of one variable to influence the change score of another variable.

Specifically, we model the change scores as

$$dy[t]_n = ys_n + \beta_y \times y[t-1]_n + \gamma_y x[t-1]_n$$
$$dx[t]_n = xs_n + \beta_x \times x[t-1]_n + \gamma_x y[t-1]_n$$

where γ_y and γ_x are called coupling parameters. γ_y represents the effect of x on the change score of y and γ_x represents the effect of y on the change score of x. We let $x0$ be the initial latent score and xs be the constant change for x. A multivariate normal distribution is assumed for the initial latent scores and constant changes for the two variables such that

$$
\begin{pmatrix} y0_n \\ ys_n \\ x0_n \\ xs_n \end{pmatrix} \sim MN \left[\begin{pmatrix} my0 \\ mys \\ mx0 \\ mxs \end{pmatrix}, \begin{pmatrix} vary0 & vary0ys & varx0y0 & vary0xs \\ vary0ys & varys & varx0ys & varxsys \\ varx0y0 & varx0ys & varx0 & varx0xs \\ vary0xs & varxsys & varx0xs & varxs \end{pmatrix} \right].
$$

Using a path diagram, a bivariate LCSM is portrayed in Figure 18.2.1.

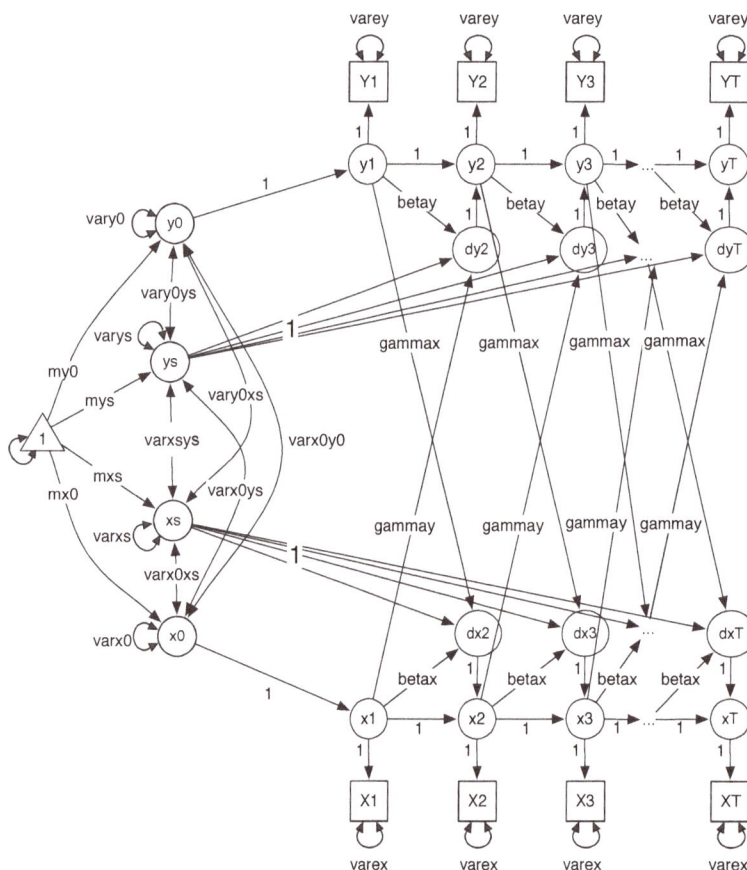

Figure 18.2.1: The path diagram for a bivariate latent change score model.

18.3   *Conducting Power Analysis for Latent Change Score Models*

18.3.1   *R package*

There are three functions in the package WebPower for power analysis for LCSM: `wp.lcsm`, `wp.blcsm`, `plot`.

The function `wp.lcsm` is used to conduct power analysis for univariate LCSMs. The basic usage of the function is given below:

```
wp.lcsm(N = 100, T = 5, R = 1000, betay = 0, my0 = 0, mys = 0,
    varey = 1, vary0 = 1, varys = 1, vary0ys = 0, alpha = 0.05,
    ...)
```

In the function, `N` is the sample size and `T` is the number of measurement occasions. Both of them can be a single value or a vector. For example, using `N=c(100,200,500)` will calculate power for the three provided sample sizes. `R` is the number of Monte Carlo simulations used to estimate the power. A larger `R` will provide more accurate power estimation but also take longer to compute. As a rule of thumb, at least 1,000 should be used. `alpha` is the significance level for testing the hypothesis of the model parameters. The default value is 0.05.

To obtain power, the population parameter values have to be provided. These values can be decided based on literature review, pilot study, expert opinions, etc. By default, all the mean, intercept and covariance parameters are set to 0 and all the variance parameters are set at 1. Those values typically have to be changed in real power analysis. Note that the name of each parameter corresponds to that used in the path diagram in Figure 18.1.1.

The output of the R function includes 4 main pieces of information for each parameter in the model. The first is the Monte Carlo estimate (`mc.est`), the mean of the `R` sets of parameter estimates from the simulated data. Note that the Monte Carlo estimates should be close to the population parameter values used in the model. The second is the Monte Carlo standard deviation (`mc.sd`), the standard deviation of the `R` sets of parameter estimates. The third is the Monte Carlo standard error (`mc.se`), the average of the `R` sets of standard error estimates of the parameter estimates. Lastly, `mc.power` is the statistical power for each parameter.

The function `wp.blcsm` is used to conduct power analysis for bivariate LCSMs. The basic usage of the function is given below. It is the same as for the univariate LCSMs.

```
wp.blcsm(N=100, T=5, R=1000, betay=0, my0=0, mys=0, varey=1,
    vary0=1, varys=1, vary0ys=0, betax=0, mx0=0, mxs=0, varex=1,
    varx0=1, varxs=1, varx0xs=0, varx0y0=0, varx0ys=0, vary0xs=0,
     varxsys=0, gammax=0, gammay=0, alpha=0.05, ...)
```

The function plot is used to generate a power curve, which has the form plot(x, parameter, ...). The first input of the function, x, is the output from either wp.lcsm or wp.blcsm. In the input of the function for power analysis, either the sample size N or the number of occasions T should be a vector. The second input is the name of the parameter to plot its power curve. Since there are multiple parameters in an LCSM, one can generate a plot for each model parameter. The name of a parameter should match the one in wp.lcsm or wp.blcsm. This function will generate one or multiple line plots in which power is shown on the y-axis and the sample size or number of occasions is shown on the x-axis.

18.3.2 *Online interface*

We also provide a Web-based interface for power analysis for LCSMs. The URL for the univariate LCSMs is http://psychstat.org/lcsm and for the bivariate LCSMs is http://psychstat.org/blcsm.

The Web interface for the univariate LCSMs is shown in Figure 18.3.1. Since the interface is built on the R function shown earlier, it requires the same input information and gives the same output. For both sample size and number of occasions, multiple values can be provided in two ways to calculate power for each given value. We discuss this using the sample size as an example since the same method is used for the number of occasions. First, multiple sample sizes can be provided and separated by spaces. For example, inputting 100 150 200 will calculate power for the three sample sizes 100, 150 and 200. Second, a sequence of sample sizes can be generated using the method *s:e:i* with *s* denoting the starting sample size, *e* as the ending sample size, and *i* as the interval. Note that the values are separated by a colon ":". For example, 100:150:10 will generate a sequence of sample sizes: 100 110 120 130 140 and 150.

18.4 *Examples*

We now show how to carry out power analysis for both univariate and bivariate LCSMs through several examples.

If the null hypothesis is true, the Monte Carlo procedure will yield the type I error rate. For example, if the parameter $\beta_y = 0$ in the population, then the estimated power should be the same as the significance level, typically 0.05. For illustration, we set the population parameter values to those shown in the second column of Table 18.4.1. Therefore, if we conduct a power analysis based on those parameter values, we will obtain the type I error rates for betay, my0, mys and vary0ys.

Example 18.4.1: Type I error for a univariate LCSM

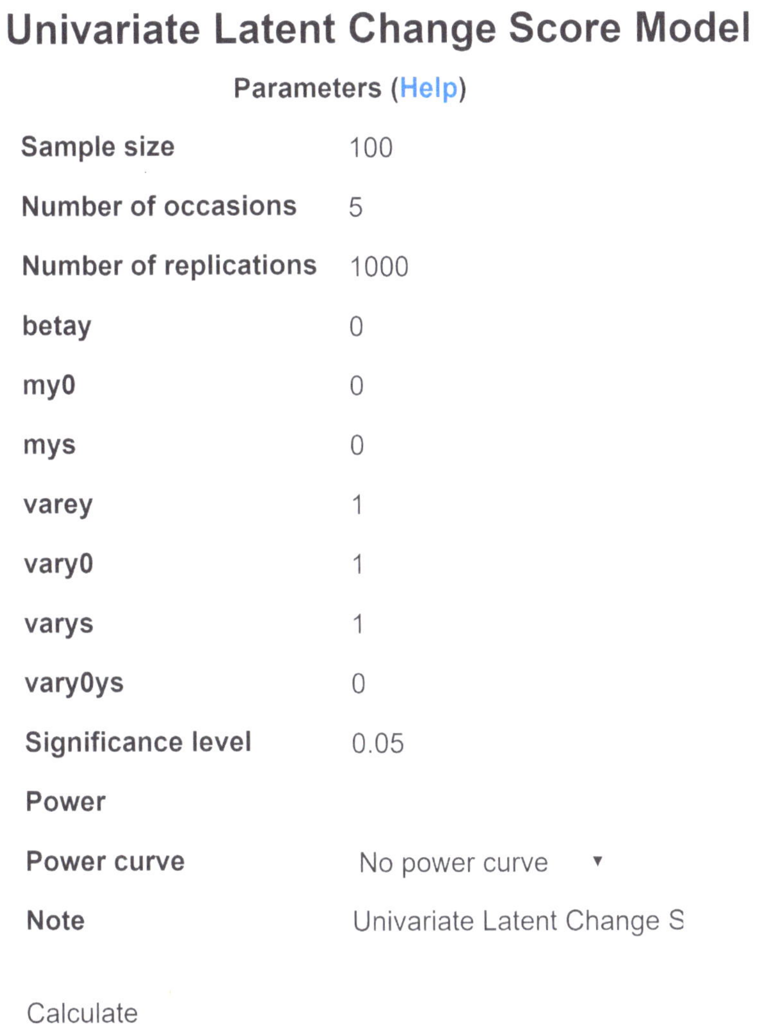

Univariate Latent Change Score Model

Figure 18.3.1: The online interface for power analysis for univariate latent change score models.

Parameters (Help)

Sample size	100
Number of occasions	5
Number of replications	1000
betay	0
my0	0
mys	0
varey	1
vary0	1
varys	1
vary0ys	0
Significance level	0.05
Power	
Power curve	No power curve ▼
Note	Univariate Latent Change S

Calculate

The R input and output for conducting the analysis are shown below. First, the estimate for each parameter is very close to the true population parameter values as shown in the column labelled mc.est. This indicates the power calculation procedure runs well. Second, the Monte Carlo standard errors are close to the corresponding Monte Carlo standard deviations, another indicator that the power calculation is trustworthy. Third, as expected, the power for betay, my0, mys, and vary0ys is close to 0.05, the nominal type I error rate. Overall, this suggests that the Monte Carlo based method can provide well-controlled type I error rate.

```
> res <- wp.lcsm(N = 100, T = 5, R = 1000, betay = 0, my0 = 0,
    mys = 0, varey = 1, vary0 = 1, varys = 1, vary0ys = 0,alpha =
    0.05)
> res
```

	Example 18.4.1	Example 18.4.2
betay	0	0.1
my0	0	20
mys	0	1.5
varey	1	9
vary0	1	2.5
varys	1	0.05
vary0ys	0	0

Table 18.4.1: Population parameter values used in Examples

```
        pop.par     mc.est   mc.sd   mc.se mc.power   N T
betay         0   0.0017944 0.05687 0.05583    0.062 100 5
my0           0  -0.0058133 0.12747 0.12555    0.051 100 5
mys           0   0.0006548 0.10283 0.10491    0.054 100 5
varey         1   0.9944460 0.08107 0.08120    1.000 100 5
vary0         1   0.9841522 0.23257 0.22956    1.000 100 5
vary0ys       0  -0.0045323 0.13427 0.13607    0.037 100 5
varys         1   0.9999032 0.22491 0.22618    1.000 100 5
```

Zhang et al. (2015) included an example of using a univariate LCSM model to analyze the WISC data. In order to plan a future study with the sample size 100 and the number of measurement occasion 5, we use the estimates as our population parameter values. Column 3 in Table 18.4.1. shows the parameter values being used in our example.

Example 18.4.2: Power analysis for a univariate LCSM

The R input and output for conducting the analysis are given below. From the output, we can see that the power to detect the parameter betay to be significant with the sample size 100 and the number of measurement occasions 5 is about 0.664. The power for another parameter, the constant change mys, is 0.274. Since oftentimes one hopes to get a power at least 0.8, a larger sample size is needed for both parameters in this study. In addition, for studying different parameters, different sample sizes are often required.

```
> wp.lcsm(N = 100, T = 5, R = 1000, betay = 0.1, my0 = 20, mys =
   1.5, varey = 9, vary0 = 2.5, varys = .05, vary0ys = 0, alpha
   = 0.05)

        pop.par    mc.est   mc.sd   mc.se mc.power   N T
betay      0.10   0.10269 0.04425 0.04363    0.665 100 5
my0       20.00  20.01389 0.31495 0.31906    1.000 100 5
mys        1.50   1.42893 1.14013 1.12052    0.284 100 5
varey      9.00   8.93260 0.72561 0.72934    1.000 100 5
vary0      2.50   2.52038 1.14866 1.14455    0.604 100 5
vary0ys    0.00  -0.03072 0.41406 0.40477    0.052 100 5
varys      0.05   0.06324 0.17373 0.17594    0.044 100 5
```

Example 18.4.2 showed that a larger sample size is needed to obtain sufficient power for parameters betay and mys. To find what sample

Example 18.4.3: Power curve for different sample sizes for a univariate LCSM

size is required, we can generate a power curve with multiple sample sizes. The R input and output for the analysis are given below. Note that seq(100, 200, 10) generates a sequence of sample sizes, and in the output power for each sample size is provided. In the plot function, we refer to a specific parameter using its name. Figure 18.4.1 shows the power curves for the two parameters betay and mys with sample sizes ranging from 100 to 200 with an interval 10. From the plot, we can easily see that to get a power 0.8 for the parameter betay, a sample size about 150 is needed. On the other hand, a sample size larger than 200 is needed for the parameter mys to have a power 0.8, with the exact number undecided based on the plot.

```
> res <- wp.lcsm(N = seq(100, 200, 10), T = 5, R = 1000,
    betay = 0.1, my0 = 20, mys = 1.5, varey = 9, vary0 =
    2.5, varys = .05, vary0ys = 0, alpha = 0.05)
> res


>
> plot(res, 'betay')
> plot(res, 'mys')
```

$'N100-T5'

	pop.par	mc.est	mc.sd	mc.se	mc.power	N	T
betay	0.10	0.100	0.044	0.044	0.627	100	5
my0	20.00	20.002	0.331	0.319	1.000	100	5
mys	1.50	1.505	1.136	1.119	0.287	100	5
varey	9.00	8.970	0.744	0.732	1.000	100	5
vary0	2.50	2.489	1.218	1.146	0.599	100	5
vary0ys	0.00	-0.009	0.413	0.403	0.059	100	5
varys	0.05	0.054	0.176	0.175	0.050	100	5

....

$'N200-T5'

	pop.par	mc.est	mc.sd	mc.se	mc.power	N	T
betay	0.10	0.100	0.031	0.031	0.915	200	5
my0	20.00	20.002	0.225	0.226	1.000	200	5
mys	1.50	1.505	0.790	0.791	0.487	200	5
varey	9.00	8.971	0.532	0.518	1.000	200	5
vary0	2.50	2.480	0.803	0.808	0.904	200	5
vary0ys	0.00	0.005	0.283	0.283	0.049	200	5
varys	0.05	0.051	0.125	0.122	0.054	200	5

Example 18.4.4: Power analysis for a bi-variate LCSM

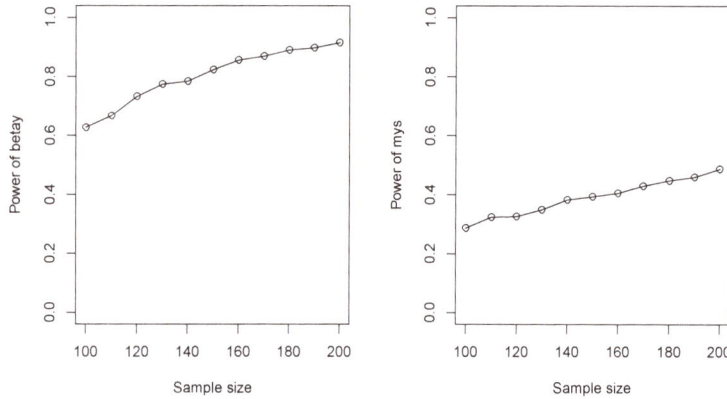

Figure 18.4.1: Power curve for betay and mys along with the sample size in the univariate latent change score model

Power analysis can be similarly conducted for bivariate LCSMs. As an example, we use the parameter estimates from a bivariate latent change score model in Zhang et al. (2015) with some modification as the population parameter values (see Table 18.4.2).

Parameter	value	Parameter	value
betay	0.08	betax	0.2
gammax	0	gammay	−0.1
my0	20	mx0	20
Mys	1.5	mxs	5
varey	9	varex	9
vary0	3	varx0	3
varys	0.05	varxs	0.6
vary0ys	0	varx0xs	0
varx0y0	1		
vayx0ys	0		
vary0xs	0		
varxsys	0		

Table 18.4.2: Population parameter values used in Example 18.4.4

The R input and output for the power analysis for a bivariate LCSM with the sample size 500 are given below. For example, for the coupling parameters gammax and gammay, the power (type I error for gammax) is 0.057 and 0.271, respectively.

```
> wp.blcsm(N=500, T=5, R=1000, betay=0.08, my0=20, mys=1.5, varey
    =9, vary0=3, varys=1, vary0ys=0, alpha=0.05, betax=0.2, mx0
    =20, mxs=5, varex=9, varx0=3, varxs=1, varx0xs=0, varx0y0=1,
    varx0ys=0, vary0xs=0, varxsys=0, gammax=0, gammay=-.1)


        pop.par    mc.est   mc.sd   mc.se mc.power   N T
betax      0.20  0.199522 0.03010 0.03066    1.000 500 5
betay      0.08  0.083244 0.06676 0.06808    0.195 500 5
```

```
gammax       0.00 -0.001327 0.02854 0.02878    0.053 500 5
gammay      -0.10 -0.098653 0.07013 0.07251    0.249 500 5
mx0         20.00 20.001958 0.14185 0.14549    1.000 500 5
mxs          5.00  4.979377 0.90537 0.94072    1.000 500 5
my0         20.00 19.998072 0.14744 0.14645    1.000 500 5
mys          1.50  1.461848 0.85601 0.88326    0.411 500 5
varex        9.00  8.994425 0.33337 0.32861    1.000 500 5
varey        9.00  8.991055 0.31327 0.32810    1.000 500 5
varx0        3.00  3.002250 0.53695 0.52353    1.000 500 5
varx0xs      0.00 -0.017027 0.23050 0.23004    0.055 500 5
varx0y0      1.00  1.014856 0.36635 0.36125    0.808 500 5
varx0ys      0.00 -0.003731 0.19780 0.20088    0.041 500 5
varxs        1.00  1.029183 0.18922 0.19046    1.000 500 5
varxsys      0.00  0.001206 0.15489 0.16238    0.043 500 5
vary0        3.00  3.027044 0.54374 0.54995    1.000 500 5
vary0xs      0.00 -0.012093 0.29175 0.29602    0.048 500 5
vary0ys      0.00 -0.025796 0.25551 0.25791    0.049 500 5
varys        1.00  1.021738 0.24698 0.25133    1.000 500 5
```

18.5   Exercises

1. Using the parameter values of Example 18.4.2, compare the statistical power of the following two conditions:

   (a) The sample size is 200 and the number of occasions is 5.

   (b) The sample size is 250 and the number of occasions is 4.

2. Using the parameter values of Example 18.4.4, generate a power curve for the coupling parameters with the sample size ranging from 500 to 1000.

Part III

Statistical Power Analysis based on Path Diagrams

19 *Drawing Path Diagrams*

Zhiyong Zhang and Yujiao Mai
Department of Psychology
University of Notre Dame

Structural equation modeling (SEM) is a statistical technique that can be used to evaluate relations among observed and latent variables (Hoyle, 1995). An SEM model typically consists of observed and latent variables where an observed variable can be measured directly but a latent variable has to be assessed using observed variables. The relations between two variables can be non-directional as correlation relationship or directional as regression relationship (Hoyle, 1995; Kline, 2011). SEM generalizes many commonly used statistical models such as confirmatory factor models, path models, latent regression models, and growth curve models (Kline, 2011). Over the past few decades, the use of SEM has been rapidly growing in many disciplines (Kline, 2011; Nachtigall et al., 2003; Westland, 2015), especially in education (Khine, 2013), psychology (MacCallum & Austin, 2000), management (Shook et al., 2004), and marketing (Babin et al., 2008), benefiting from the fact that an SEM can be represented graphically as a path diagram.

19.1 *Traditional Path Diagrams*

Figure 19.1.1 shows a typical path diagram for SEM. An SEM path diagram consists of two types of graphical components: variables and paths. The variables are also called vertices/nodes and the paths are also called edges in graph theory.

19.1.1 *Variables*

Three types of variables are allowed in a regular SEM path diagram: observed, latent and constant variables. An observed variable is represented by a rectangle, a latent variable is represented by an ellipse node, and a constant variable is represented by a triangle. The constant variable is a special variable with value 1. It is used to represent a mean or an intercept.

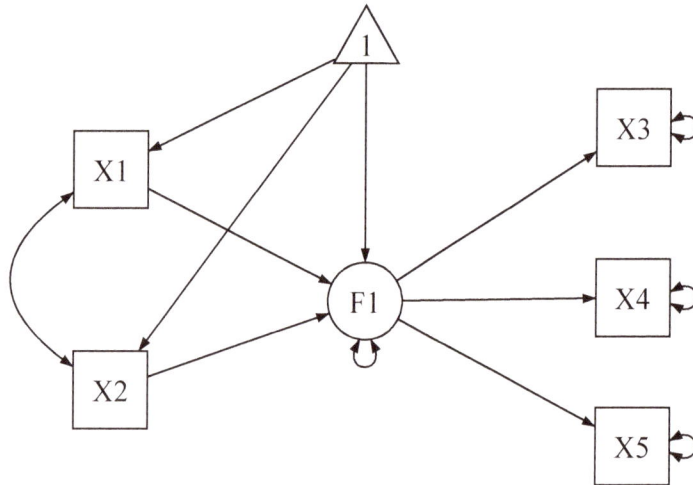

Figure 19.1.1: A typical SEM path diagram example

19.1.2 Paths

Two types of paths, a directed one and an undirected one, are allowed to define the relationship between the variables in a model or diagram. A directed path, represented by a single-headed arrow, is used to indicate that one variable can predict another variable. It usually means a regression relationship. A directed path points from the predicting variable toward the outcome variable. More specifically in SEM, a directed path from an ellipse (a latent variable) to a rectangle (an observed variable) represents a factor loading, otherwise, it represents a regression coefficient.

An undirected path is represented by a double-headed arrow. If an undirected path connects two variables, it is the covariance between the two variables. An undirected path can start from one variable and end on the same variable. Such a path can represent either a variance or a residual variance in a structural equation model. For a variable with such a path, if there is no directed path pointing to it, the path represents a variance; otherwise, a residual variance.

Both variables and paths can be labeled or named. The labels should be as meaningful as possible. A number on a path is simply the value of the coefficient represented by the path.

19.1.3 Basic rules for drawing SEM path diagram

To ensure consistency and improve usability, the following rules are imposed in constructing path diagrams. First, no more than one triangle variable can be created in one diagram. Second, if existing, a directed path can only be drawn starting from the triangle variable. Third, either a directed or an undirected path can be drawn between an observed variable and a latent variable. Fourth, an undirected path (double-headed arrow) is automatically created for a variable that is predicted by one or more variables. The path is also automatically removed when the involved variable disconnects from all its predicting variables. Finally, when a variable is removed, all paths connecting to it will be removed automatically.

19.2   *Path Diagram in WebPower*

WebPower can conduct statistical power analysis by drawing a path diagram directly online. The path diagram application is developed using JavaScript with the library d3js (Myatt & Johnson, 2011) and the library jQuery (Resig et al., 2009). The path diagrams can be saved in the format of scalable vector graphics (SVG) (Ferraiolo et al., 2000). In addition to the variables and paths used in the regular path diagrams, we also provide the support of non-normal data, missing data, and multilevel model representations as we will discuss later.

19.2.1   *Interface and buttons*

The overall path diagram interface is shown in Figure 19.2.1. Intuitively, a path diagram is constructed using the buttons on the left panel.

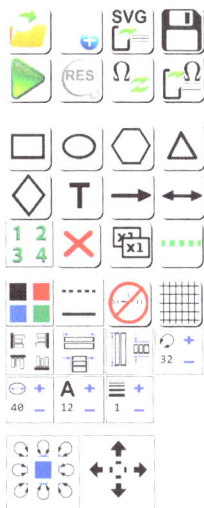

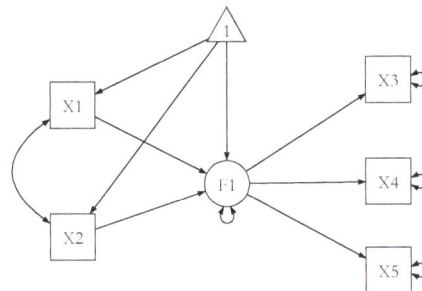

Figure 19.2.1: WebPower path diagram interface

The table displays the buttons in the diagram and their functionalities.

Button	Functionalities

Open an existing path diagram

Start a new diagram

Save a diagram as a SVG image

Save a diagram

Run power analysis based on a diagram

View the power analysis result

Process LaTeX equations

Save a SVG diagram with LaTeX equations

Add a square to the canvas, representing an observed variable.

Add a circle to the canvas, representing a latent variable.

Add a hexagon to the canvas. It should have the same name as an observed variable since it is used to specify how data in the variable are missing.

Add a triangle to the canvas, representing either a mean or an intercept.

Add a diamond to the canvas. It is used to specify the non-normality of data. On the path, the skewness and kurtosis for a variable can be specified.

Add any text to the canvas.

Draw a single-headed arrow between two shapes using one of the two ways. Firstly, one can start by clicking the line button, then drag from the start node to the end node. Secondly, one can first select the start node, then click the line button, and finally click the end node.

Draw a double-headed arrow between two shapes using one of the two ways. Firstly, one can start by clicking the line button, then drag from the start node to the end node. Secondly, one can first select the start node, then click the line button, and finally click the end node.

Set the level of variables. Each number can be clicked as a separated button to set a variable to be a 1st, 2nd, 3rd or 4th level variable.

Delete a selected shape from the canvas.

Copy selected shape(s). To select multiple shapes, drag the mouse around the shapes.

Add a line to the canvas.

Change the color of the selected shapes.

Change the lines or shapes from solid to dashed or vice versa.

Hide or show the labels on the paths.

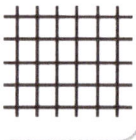

Hide or show the grid lines on the canvas.

Align selected shapes according to the sides.

Resize the selected shapes to have the same width.

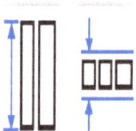

Resize the selected shapes to have the same height.

Change the size of the double-headed arrow on a variable. Clicking "+" to increase and "-" to decrease the size. The size as a number is shown on the bottom left corner. One can also change the size by inputting a number and then clicking the icon on the top-left corner to apply the modification.

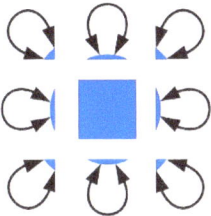

Change the size of the selected shapes. Clicking "+" to increase and "-" to decrease the size. The size as a number is shown on the bottom left corner. One can also change the size by inputting a number and then clicking the icon on the top-left corner to apply the modification.

Change the size of the text. Clicking "+" to increase and "-" to decrease the size. The size as a number is shown on the bottom left corner. One can also change the size by inputting a number and then clicking the icon on the top-left corner to apply the modification.

Change the size of the paths. Clicking "+" to increase and "-" to decrease the size. The size as a number is shown on the bottom left corner. One can also change the size by inputting a number and then clicking the icon on the top-left corner to apply the modification.

Position the double-headed arrow on a variable. Eight different positions are allowed.

Move shapes on a canvas around. The four arrows tell the direction to move. One can also move a shape by dragging it using the mouse.

19.2.2 *How to draw a path diagram*

Using the buttons on the interface, one can draw a new or edit an existing path diagram interactively.

- To create a new diagram, click the button (⬢) and to edit an existing diagram, click the button (📁) to first open it.

- A rectangle/square variable (▢) or an ellipse/circle variable (◯) can be created by clicking their corresponding buttons in the interface. A hexagon (⬡) can also be created in the same way.

- A new path can be drawn in two ways. Firstly, one can start by clicking the path button, then drag from the start variable to the end variable. Secondly, one can first select the start variable, then click the path button, and finally click the end variable. Both the single-headed (→) and double-headed (↔) arrows can be drawn in this way. For the double-headed arrow on a variable itself, the start variable and the end variable are the same. The position of the double-headed arrow on a variable itself can be changed by dragging it around or use this button (⬚).

- A line can be changed between a solid line and a dotted line using the button (—).

- To create a triangle shape (△), one needs to first select a rectangle, ellipse, or a hexagon shape and then click on the triangle button. A single-headed path will also be added automatically from the triangle shape to the shape selected.

- Both shapes and paths can be duplicated easily. To duplicate one shape, simply select it by clicking it and then click the copy button (⬚). To copy and paste multiple shapes, one can select all of them together first.

- Select multiple items on the canvas. There are two ways to do it. First, one can hold the "Ctrl" button on the keyboard and then click the items to be selected. Second, one can hold the left button of the mouse and drag a square on the canvas to include all the items to be selected.

- Both shapes and paths can be named. To name a shape or path, first double-click it and then change or add a name in the pop-up dialog window as shown below.

Edit Text

X1

OK Cancel

- To add text to the canvas, click the button (**T**) and then type in the text in the pop-up dialog window.

- To draw a line anywhere on the canvas, click the button ( ).

- Everything within a path diagram can be moved freely by one of the two ways. Firstly, one can simply drag it around. Secondly, one can first select a shape and then use the buttons ( ) on the left-hand panel to move it around. To move multiple shapes, one can select all of them and then drag them to the desired location.

- The properties of a selected shape or path can be modified by clicking the property buttons such as color ( ), font size ($\overset{A}{_{12}}$ +) and stroke width ($\overset{\equiv}{_{1}}$ +). Clicking "+" to increase and "-" to decrease the size. The size as a number is shown on the bottom left corner. One can also change the size by inputting a number and then clicking the icon on the top-left corner to apply the modification.

- To delete a shape or path, simply select it and then click the delete button (✕).

- One can align multiple selected shapes on their four sides by using the button ( ).

- To set the selected shapes to have the same width using ( ) and same height using ( ).

- **LaTeX equation**. Math formulas can be put anywhere on the canvas through LaTeX notation. First, type in the equation as you would do in LaTeX, e.g., α for α, and $\alpha^2 +\beta^2$ for $\alpha^2 + \beta^2$. Then click the button ($\overset{\Omega}{\rightleftharpoons}$) to show the equations. Note that when clicking the canvas, the equation will go back to text. To save the diagram with the math equations, click the button ( ). The path diagram will be saved into an HTML file that can be viewed in a web browser.

- To save the path diagram on the server to use in the future, click the button ( ). The saved diagram can be opened for edits in the future.

- The diagram can also be saved to local storage by clicking the button ( ). This saves the diagram into an SVG figure that can be used in publication or converted to other format such as PDF, JPEG, or PNG.

20 Power Analysis based on Monte Carlo Simulation through Diagrams

Zhiyong Zhang and Yujiao Mai
Department of Psychology
University of Notre Dame

Except for some statistical tests, it is generally difficult to get the analytically traceable distribution for a statistic under the alternative hypothesis even though it is possible to get its distribution under the null hypothesis. Therefore, statistical power analysis for complex models has to rely on Monte Carlo simulation. The typical Monte Carlo based power analysis method assumes that the sampling distribution of a statistic is known under the null hypothesis, which we refer to as regular Monte Carlo method. It is also possible to conduct power analysis while the sampling distribution is not known under the null hypothesis, which we refer to as a double Monte Carlo method in this chapter (Yuan et al., 2017). Both methods are implemented in WebPower. We illustrate how to use them in this chapter.

20.1 A Regular Monte Carlo Based Method

20.1.1 Basic idea

For a given statistics T, by definition, its power

$$\pi = \Pr(\text{reject } H_0 | H_1) = \Pr(T > c_{1-\alpha} | H_1)$$

where H_0 and H_1 are null and alternative hypothesis, respectively, and $c_{1-\alpha}$ is a critical value from the sampling distribution of T under the null hypothesis with α denoting the significance level. The regular Monte Carlo based method works when the critical value $c_{1-\alpha}$ can be obtained. Then we can approximate the power using the relative frequency to reject the null hypothesis given the alternative hypothesis is true so that

$$\hat{\pi} = \frac{1}{R} \sum_{i=1}^{R} (\hat{T}_i > c_{1-\alpha})$$

where R is the total number of Monte Carlo simulation and $\hat{T}_i$ is the sample statistic under the ith replication. More specifically, the following procedure can be used.

1. Decide the significance level. Usually, the default 0.05 can be used. Based on that, get the critical value $c_{1-\alpha}$.

2. Specify a model with the hypothesized population parameter values (θ).

3. Generate a set of data with the sample size N from the model using random number generation techniques.

4. Fit the hypothesized model to the generated data and obtain the statistic $\hat{T}$.

5. If $\hat{T} > c_{1-\alpha}$, the null hypothesis H_0 is rejected.

6. Repeat Steps (2)–(5) for a total of $R (R \geq 1000)$ times.

7. Suppose out of the R replications, the null hypothesis H_0 is rejected r times. Then the statistical power with the sample size N is estimated by $\hat{\pi} = \frac{r}{R}$.

8. For sample size planning, if $\hat{\pi}$ is smaller than the desired power, say 0.8, one can increase the sample size to repeat Steps 2 and 7 to recalculate the power. Otherwise, the sample size can be set to a smaller value.

We discuss several issues regarding the Monte Carlo based method.

Critical value $c_{1-\alpha}$. In SEM and multilevel modeling, when testing individual parameter in the model, it is often assumed that a parameter follows a normal distribution under the null hypothesis. Therefore, in WebPower, we decide the critical value based on normal distribution assumption. For example, if the significance level $\alpha = 0.05$ is used, the critical value $c_{0.95} = 1.96$ is used based on the normal distribution function.

Multiple parameters. Although a single parameter from a model might be the focus of power analysis, using the Monte Carlo based procedure, one can obtain the power for every parameter in the model under investigation. Therefore, WebPower outputs the power for each parameter in a model.

Missing data. The influence of missing data on power can be evaluated by specifying how missing data are generated. To do so, add a (⬡) to the diagram and give it the same name as the variable with missing data. Then, the missingness can be specified in a logistic regression way as shown in the example below.

Non-normal data. The influence of non-normal data on power can be evaluated. To do so, add a (◇) to the diagram. First select the non-normal variable, and then click (◇). On the path from the diamond to the non-normal variable, one can input skewness and kurtosis for the variable. The values for skewness and kurtosis are separated by ";".

20.1.2 *Examples*

We now illustrate the Monte Carlo based power analysis using the path diagram interface of WebPower.

http://psychstat.org/diagram

Figure 20.1.1 shows the path diagram and other information for conducting power analysis for a simple mediation model. In this model, X, M, and Y are the predictor, mediator, and output, respectively. A label on the path represents the population value of the parameter.

Example 20.1.1: Power for a simple mediation effect in the simple mediation model

Sample Size:

100

Significance level:

.05

MC replications:

100

Notes:

MC simulation using dia

Power parameters:

ab := a*b

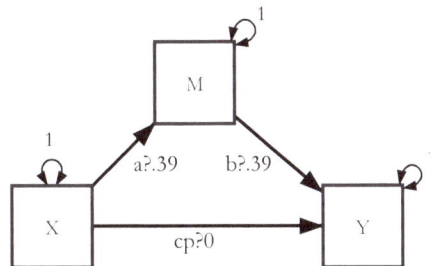

Figure 20.1.1: A simple mediation model example

For a given path, different labels can be used. A label can consist of letters, numerical values and one of the symbols: "@", "?" and ";". The general rules are given below.

- No label or a text label: the population parameter for the path is fixed at 0 for single-headed arrows. For double-headed arrow, if it is a variance, it is fixed at 1, otherwise, 0 for covariance. A text label can consist of letters and numbers but has to start with a letter. For example, "p1" can be a text label but "1p" cannot.

- A numerical value: the population parameter takes the numerical value.

- A symbol "@" or "?" and a numerical value such as @1 or ?1. The numerical value is set as the population value in the model. If it is preceded by "@", when estimating the model, the parameter is set to be a fixed one at the value. If it is preceded by "?", the parameter is freely estimated.

- A text label + either "@" or "?" + a numerical value such as p1@0.5 or p2?0.3. This is similar to the case of a symbol with a numerical value. However, the path is given a name as the text and can be used in other situation.

- If multiple paths are given the same label (name), the corresponding parameters are constrained to be equal in estimation.

- The symbol ";" is typically used in multiple group analysis. For example, a1;a2@.5;.3 means that for the first group, a1 fixed at 0.5 and for the second group a2 fixed at 0.3 when estimating the model. b1;b2?0;0 set the population values at 0 and the corresponding parameters will be estimated for both groups. d1;d1 means the parameters for both groups are set to be the same when estimating the model.

Therefore, for the mediation model, the effect of X on M is 0.39 and the parameter is called a; and the effect of M on Y after controlling X is also 0.39 and the parameter is called b. The direct effect is 0 with the name cp. All the variance parameters are set at 1. All the parameters will be freely estimated when calculating statistical power.

As with any power calculation, we also need to provide some basic information. For example, the sample size under evaluation is always required. Here, we evaluate power for a sample size 100 as specified in the field called Sample Size on the left panel below all the buttons. Furthermore, we also need to specify the significance level, 0.05 by default, and the number of Monte Carlo replications (MC replications), with 1,000 as the default.

Typically, power is estimated for every parameter in a model. For a mediation effect, it is a combination of multiple parameters. Therefore, to calculate power for it, one has to construct a new parameter. The new parameter is constructed using the symbol ":=" where the new parameter is on the left, and on the right regular mathematical operations can be used. For this specific example, we are interested in the mediation effect via M. So we define the new parameter ab using ab := a*b. The newly defined parameter is put in the field "Power parameters". Each line in this field can represent a newly defined parameter.

Once the path diagram is created with all the information, clicking the button will initialize the power calculation on the server.

Clicking the button (RES) will open the output window. Be patient as the calculation takes a relatively long time due to the involvement of Monte Carlo simulation. For the simple mediation example, the calculation took about 139 seconds based on 1,000 Monte Carlo replications.

The output for the analysis is in Figure 20.1.2. The main result is shown in the middle of the output. In addition, it provides the time used to run the analysis. The path diagram used to conduct the analysis can be viewed and edited for a new power analysis by clicking the button at the bottom.

WebPower started at 10:05:33 on Mar 06, 2017.
=====================================
Please refresh your browser if you are expecting more output.

Figure 20.1.2: Complete output of the simple mediation example

```
Basic information:

  Esimation method                        ML
  Standard error                    standard
  Number of requested replications      1000
  Number of successful replications     1000
  Sample size                            100

                 True  Estimate   MSE     SD    Power Power.se Coverage
Regressions:
  M ~
    X        (a)   0.390   0.389   0.100   0.101  0.969   0.005   0.950
  Y ~
    M        (b)   0.390   0.387   0.100   0.105  0.961   0.006   0.934
    X        (cp)  0.000   0.005   0.108   0.110  0.060   0.008   0.940

Intercepts:
    M             0.000  -0.001   0.099   0.099  0.045   0.007   0.955
    Y             0.000   0.003   0.099   0.101  0.059   0.007   0.941
    X             0.000  -0.003   0.099   0.099  0.059   0.007   0.941

Variances:
    M             1.000   0.979   0.138   0.143  1.000   0.000   0.926
    Y             1.000   0.970   0.137   0.143  1.000   0.000   0.895
    X             1.000   0.986   0.139   0.143  1.000   0.000   0.928

Indirect/Mediation effects:
    ab            0.152   0.151   0.056   0.059  0.872   0.011   0.919
```

=====================================
WebPower ended at 10:07:49 on Mar 06, 2017

Time spent on the analysis

```
   user  system elapsed
 135.990   0.545 136.539
```

To see the input diagram:

The results of the analysis consist of the following information.

- The "Estimation method" is used to estimate the model in each replication.

- The "Standard error" estimation method for the parameter estimates, can be the regular normal based method or the robust method.

- The "Number of requested replications" and the "Number of successful replications".

- The "Sample size" provides the sample size used in the analysis.

- The column "True" lists the population values of parameters used in the simulation.

- The column "Estimate" is calculated as the mean of the parameter estimates from the simulated data based on the "Number of successful replications". Note that the Monte Carlo estimates should be close to the population parameter values used in the model.

- The column "MSE", mean standard error, is the average of the standard error estimates of the parameter estimates based on the "Number of successful replications".

- The column "SD", empirical standard deviation, is calculated as the standard deviation of the parameter estimates from the successful replications.

- The column "Power" provides the statistical power for each parameter.

- The column "Power.se" provides the standard error of the power estimate. It is simply calculated as

$$Power.se = \sqrt{\frac{\pi(1 - \pi)}{R}},$$

where π is the estimated power and R is the number of successful replications.

- The column "Coverage" provides the coverage rate of confidence interval constructed based on the estimated standard error and normal approximation.

In the result, the information for each individual parameter is first provided on each row. Then the newly defined parameters are listed under the heading of "Indirect/Mediation effects". Specifically for this example, the power to detect the mediation effect is about 0.878 with a sample size 100.

Statistical power analysis can be conducted by taking into account of both non-normal data and missing data. Figure 20.1.3 shows a mediation model where the outcome variable F1 is a latent variable measured by X3, X4 and X5. Both X1 and X2 are non-normal variables. The variable X5 has missing data and the missingness is related to X1.

Currently, only continuous data with skewness and kurtosis different from normal distribution can be used. To specify the skewness and kurtosis for a given variable, draw a path from the diamond ◇ to the variable. Then, on the path, use the label with two values separated by ";". The first value is the skewness and the second is kurtosis.

Example 20.1.2: Power for a mediation effect with latent variable, non-normal data and missing data

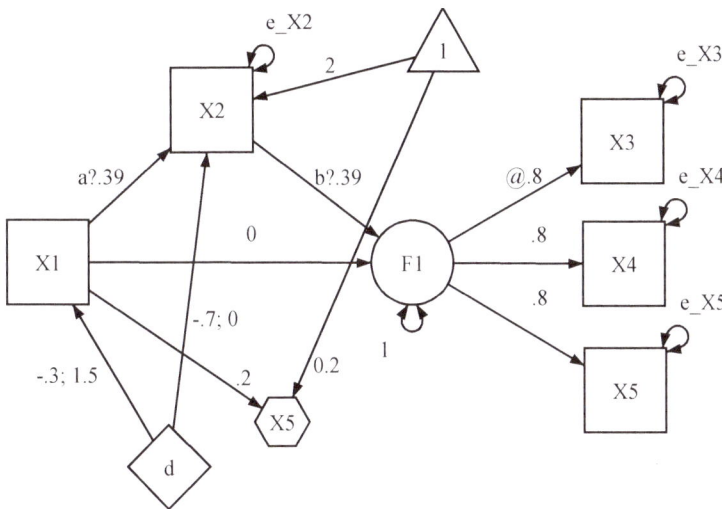

Figure 20.1.3: Mediation model with non-normal and missing data

Missing data are specified through logistic models. Current, only missing completely at random (MCAR) and missing at random (MAR) mechanisms are supported. To specify how missing data are generated, first draw a hexagon (⬡) on the canvas and give it the same name as the variable with missing data. Note that missing data are only allowed in the manifest variables. Our software will generate missing data according to the logistic model

$$\log \frac{p_i}{1 - p_i} = \beta_0 + \beta_1 x_{1i} + \ldots + \beta_k x_{ki}$$

where p_i is the probability for a datum to be missing in the observed data and x's are the variables related to the missingness. Note that if $\beta_1 = \ldots = \beta_k = 0$, the missing probability is a constant and

$$p_i = \frac{\exp(\beta_0)}{1 + \exp(\beta_0)}.$$

This also indicates that the missing mechanism is MCAR. With co-variates in the MAR case, the missing probability can be calculated similarly.

The output for the power analysis based on the model in Figure 20.1.3 is shown in Figure 20.1.4. Note that since there are non-normal data, the robust procedure (robust.huber.white) is used to estimate the standard errors for hypothesis testing. From the output, the power for detecting the mediation effect is about 0.582.

```
Basic information:

  Esimation method                              ML
  Standard error                   robust.huber.white
  Number of requested replications            1000
  Number of successful replications           1000
  Sample size                                  100

                     True  Estimate    MSE      SD    Power Power.se Coverage
Latent variables:
  F1 =~
    X3             0.800     0.800   0.000   0.000    NaN     NaN    0.000
    X4             0.800     1.831   0.315  31.273  0.896   0.010    0.934
    X5             0.800     0.827   0.293   0.298  0.882   0.010    0.935

Regressions:
  X2 ~
    X1       (a)   0.390     0.395   0.099   0.102  0.971   0.005    0.926
  F1 ~
    X2       (b)   0.390     0.393   0.149   0.158  0.757   0.014    0.918
    X1             0.000    -0.002   0.149   0.146  0.053   0.007    0.947

Intercepts:
    X2             2.000     2.003   0.099   0.100  1.000   0.000    0.936
    X3             0.000    -0.008   0.270   0.280  0.069   0.008    0.931
    X4             0.000     0.000   0.270   0.275  0.080   0.009    0.920
    X5             0.000    -0.009   0.321   0.341  0.081   0.009    0.919
    F1             0.000     0.000   0.000   0.000    NaN     NaN    0.000

Variances:
    X2    (e_X2)   1.000     0.981   0.138   0.139  1.000   0.000    0.913
    F1             1.000     1.029   0.515   0.497  0.767   0.013    0.915
    X3    (e_X3)   1.000     0.939   0.334   0.309  0.880   0.010    0.955
    X4    (e_X4)   1.000     0.451   0.302  16.042  0.877   0.010    0.938
    X5    (e_X5)   1.000     0.921   0.319   0.324  0.851   0.011    0.903

Indirect/Mediation effects:
    ab                       0.152   0.154   0.072  0.075  0.582   0.016   0.914
```

Figure 20.1.4: Power analysis results for mediation model with non-normal and missing data

Power analysis can also be conducted for multilevel models. Figure 20.1.5 shows the path diagram for a two-level model. In its mathematical format, the model can be written as:

Example 20.1.3: Power for a two-level model

$$\text{Level 1}: \quad Y1_{ij} = b_{i0} + b_{i1}X1_{ij} + e_{ij}$$
$$\text{Level 2}: \quad \begin{aligned} b_{i0} &= \beta_1 + \beta_2 Z1_i + v_{i0} \\ b_{i1} &= \beta_3 + \beta_4 Z1_i + v_{i1} \end{aligned} \quad .$$

Note that in the path diagram, the random-effects b_{i0} and b_{i1} are used as labels in the first level and plotted as latent variables in the second level. Based on the path diagram, we have the population values of parameters for the fixed-effects parameters $\beta_1 = 0.3$ (the path from the triangle to b0), $\beta_3 = 0$ (the path from the triangle to b1), $\beta_2 = 0.45$ (the path from Z1 to b0) and $\beta_4 = 0.3$ (the path from Z1 to b1). For the random-effects parameters, the variance of e_{ij} is 0.6 as shown on the double-headed path related to Y1, the variance of X_{ij} is 0.9, the variance of v_{i0} is 0.8, the variance of v_{i1} is 0.9, and the covariance between v_{i0} and v_{i1} is 0. Since no variance is specified for Z, it is set at 1 by default.

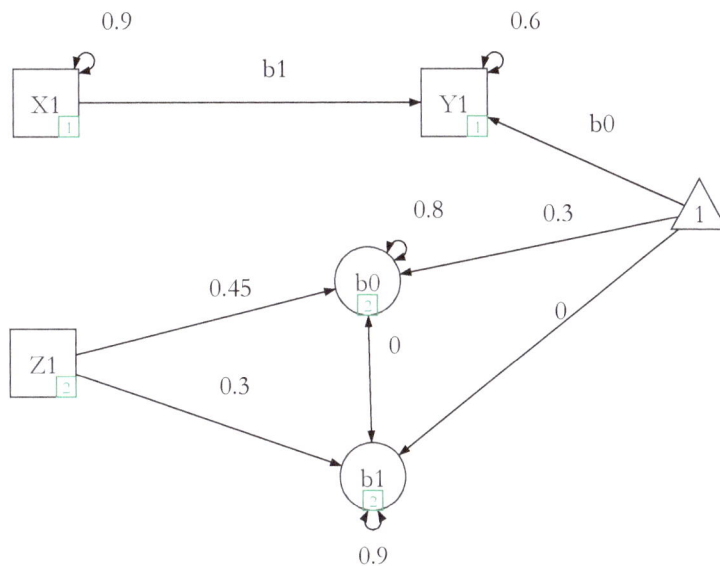

Figure 20.1.5: Path diagram for a two-level model

Based on the model input, we conduct a power analysis with 30 groups (level 2 sample size) and 100 participants in each group (level 1 sample size). This is provided in the "Sample size" field with the format 100,30. We then set the Monte Carlo replications to be 1,000. The output for the power analysis is shown in Figure 20.1.6.

For the multilevel models, only power for the fixed-effects parameters is produced because these parameters are often of interest. Currently, lmer package used by WebPower does not provide standard errors for random-effects parameters and therefore no power values are provided for those parameters. Second, the power is estimated based on the standard error estimates from the REML estimation method. Third, as for SEM, the power for each fixed-effects parameters is provided together with other information. Especially, to detect the effect of Z1 on b0, the random intercept, the power is about 0.74, while to detect the effect of Z1 on b1, the random slope, the power is about 0.41.

Missing data can also be considered when calculating power for

Example 20.1.4: Power for a two-level model with missing data

```
Esimation method                                    REML
Number of requested replications                    1000
Number of successful replications                   1000
Number of levels                                       2
Sample size for different levels                  100,30
```

Level	Equation	Label	True	Estimate	MSE	SD	Power	Power.se	Coverage
	Fixed effects								
2	b0~1		0.3	0.290	0.165	0.167	0.428	0.016	0.947
2	b1~1		0	-0.011	0.175	0.181	0.063	0.008	0.937
2	b0~Z1		0.45	0.441	0.170	0.175	0.740	0.014	0.942
2	b1~Z1		0.3	0.299	0.180	0.182	0.410	0.016	0.948
	Random effects								
1	Y1~~Y1		.6	0.599		0.016			
2	b0~~b0		.8	0.798		0.209			
2	b1~~b1		.9	0.898		0.241			
2	b0~~b1		0	-0.005		0.166			

Figure 20.1.6: Power analysis results for the two-level model in Figure 20.1.5.

multilevel models. To include missingness, one needs to specify the missing data mechanism. For example, in the diagram in Figure 20.1.7, we draw a hexagon named Y1 and then draw a directed path from X1 to predict it. This means that the missingness in Y1 is related to X1 as in a logistic regression

$$\text{logit}[\Pr(Y_1 \text{ is missing})] = 0.2 \times X_1.$$

Therefore, if X_1 takes a greater value, Y_1 is more likely to be unobserved during data collection.

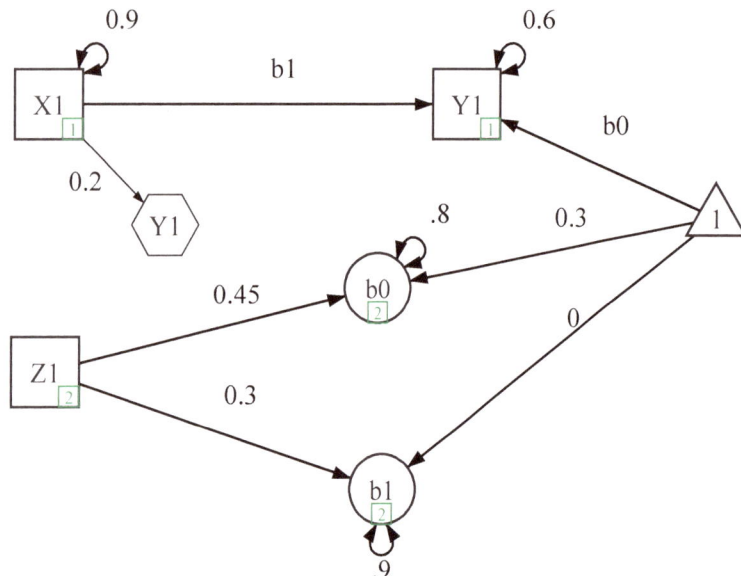

Figure 20.1.7: Path diagram for a two-level model with missing data

For the model specified in Figure 20.1.7, the power analysis is shown in Figure 20.1.8.

Non-normal data is also allowed in conducting power analysis. For

Example 20.1.5: Power for a two-level model with missing data and non-normal data

```
    Esimation method                                      REML
    Number of requested replications                      1000
    Number of successful replications                     1000
    Number of levels                                         2
    Sample size for different levels                   100,30
```

Level	Equation	Label	True	Estimate	MSE	SD	Power	Power.se	Coverage
	Fixed effects								
2	b0~1		0.3	0.503	0.167	0.162	0.851	0.011	0.759
2	b1~1		0	-0.002	0.176	0.178	0.066	0.008	0.934
2	b0~Z1		0.45	0.447	0.172	0.176	0.738	0.014	0.944
2	b1~Z1		0.3	0.301	0.182	0.196	0.399	0.015	0.930
	Random effects								
1	Y1~~Y1		.6	0.600		0.023			
2	b0~~b0		0.8	0.807		0.226			
2	b1~~b1		0.9	0.899		0.241			
2	b0~~b1		0	-0.001		0.166			

Figure 20.1.8: Power analysis results for the two-level model with missing data in Figure 20.1.7.

example, the model in Figure 20.1.9 includes a diamond shape with D on it. This indicates that Y1 is non-normally distributed with skewness 1.5 and kurtosis 0.8. By such specification, the power is calculated by simulating non-normal data for Y1. However, the typical normal-based estimation method is still used here. Therefore, the estimated power is often not reliable and we do not show the output here.

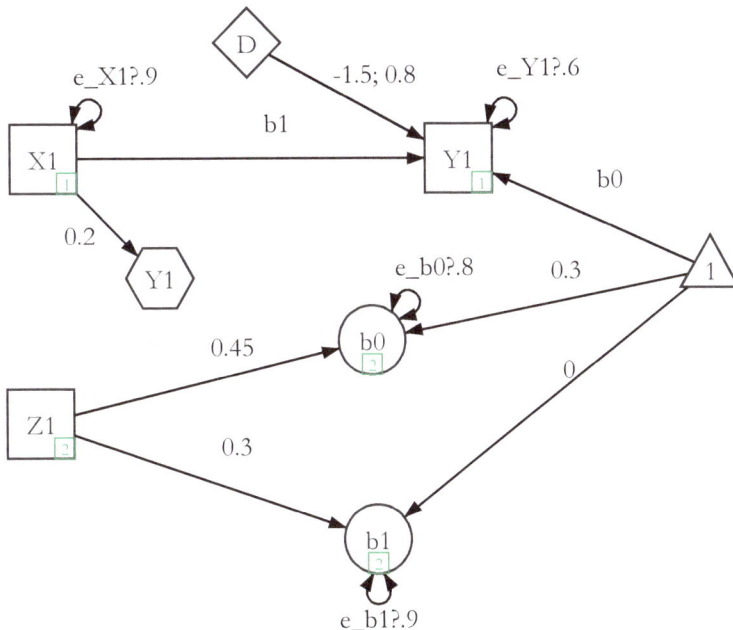

Figure 20.1.9: Power analysis results for the two-level model with missing data and non-normal data.

Power analysis can be similarly conducted for three-level models. Figure 20.1.10 shows the diagram of a three-level model, which can be

Example 20.1.6: Power for a three-level model

expressed mathematically as

$$\text{Level 1}: \quad Y1_{ijk} = b_{ik0} + b_{ik1}X1_{ijk} + e_{ijk}$$
$$\text{Level 2}: \quad \begin{aligned} b_{ik0} &= \beta_1 + d_k Z1_{ik} + v_{ik0} \\ b_{ik1} &= \beta_2 + \beta_3 Z1_{ik} + v_{ik1} \end{aligned}$$
$$\text{Level 3}: \quad d_k = \beta_4 + \beta_5 W1_k + w_k$$

Note that the relationship between Z1 and the random intercept b0 varies in the second level of the model. It is therefore specified as a latent variable in the third level. Furthermore, it is predicted by the third level predictor W1.

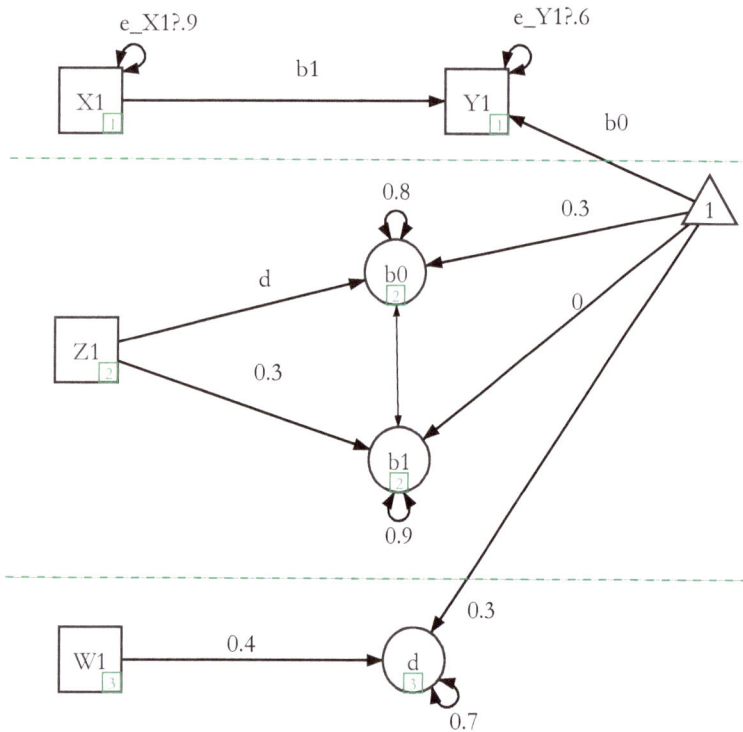

Figure 20.1.10: A three-level model example

The output for the power analysis is given in Figure 20.1.11. The output has the same format as for the two-level models.

20.2   A Double Monte Carlo Procedure for Power Analysis

The regular Monte Carl method used in the previous section 20.1 requires that the distribution of a test statistic under the null hypothesis is known. However, the null distribution can be difficult to derive when the models become more complex. For instance, in the example in Figure 20.1.9, the distribution of parameters under the null hypothesis is not clear because of missing data and non-normal data. When

```
Esimation method                                    REML
Number of requested replications                    1000
Number of successful replications                    520
Number of levels                                       3
Sample size for different levels                50,30,30
```

Level	Equation	Label	True	Estimate	MSE	SD	Power	Power.se	Coverage
	Fixed effects								
2	b0~1		0.3	0.299	0.034	0.032	1.000	0.000	0.954
2	b1~1		0	-0.000	0.032	0.032	0.048	0.009	0.952
3	d~1		0.3	0.306	0.155	0.158	0.510	0.022	0.935
3	d~W1		0.4	0.412	0.155	0.161	0.744	0.019	0.933
2	b1~Z1		0.3	0.301	0.032	0.034	1.000	0.000	0.944
	Random effects								
1	Y1~~Y1	e_Y1	.6	0.600		0.004			
2	b0~~b0		0.8	0.800		0.040			
2	b1~~b1		0.9	0.904		0.042			
2	b0~~b1		0	-0.002		0.030			
3	d~~d		0.7	0.677		0.197			

Figure 20.1.11: Power analysis results for the two-level model with missing data in Figure 20.1.10.

conducting power analysis, the population information about a model is supposed to be known. Therefore, the null distribution can be similarly simulated based on Monte Carlo methods.

Using the idea, we first estimate the empirical distribution of a parameter of interest and the desired critical value through the first Monte Carlo simulation under the null hypothesis. Then, we estimate the statistical power using the relative frequency of rejecting the null hypothesis in a second Monte Carlo simulation based on the critical values obtained in the first Monte Carlo simulation. We call the new method a double Monte Carlo procedure for power analysis.

Specifically, the following procedure can be used to obtain power for a parameter in a model. Let M_0 and M_1 be the models under the null hypothesis H_0 and the alternative hypothesis H_1, respectively.

1. Simulate the critical value $c_{1-\alpha}$

    (a) Specify Model M_0 under H_0 with hypothesized population parameter values (θ) but setting $\gamma = \gamma_0$ with γ_0 being the parameter values under the null hypothesis.

    (b) Generate Q sets of data with sample size N from M_0. Missing data and non-normal data can be generated so that their effects on the statistic T are accounted for. Q should be at least 1,000. By default, $Q = 10,000$ in WebPower.

    (c) Fit Models M_0 and M_1 to the generated data and obtain a sample test statistic $\hat{T}$. Then, Q values of the test statistic are available, which characterize the empirical distribution of T under H_0.

    (d) Get the critical value $c_{1-\alpha}$ as the $100(1 - \alpha)$th percentile of the empirical distribution of T.

2. Simulate the statistical power π

(a) Specify Model M_1 under H_1 with hypothesized population parameter values (θ).

(b) Generate a set of data with sample size N from the model using random number generation techniques. Missing data and non-normal data can be generated so that their influences on the statistic T under H_1 are accounted for.

(c) Fit Models M_0 and M_1 to the generated data and obtain the sample test statistic $\hat{T}$.

(d) If $\hat{T} > c_{1-\alpha}$, where $c_{1-\alpha}$ is from Step 1, the null hypothesis H_0 is rejected.

(e) Repeat Steps (b)-(d) for a total of $R(R \geq 1000)$ times.

(f) Suppose out of the R replications, the null hypothesis H_0 is rejected r times. Then the statistical power with the sample size N is estimated by $\hat{\pi} = r/R$.

3. For sample size planning, if $\hat{\pi}$ is smaller than the desired power, say 0.8, one can increase the sample size to repeat Steps 1 and 2 to recalculate the power. Otherwise, the sample size can be set to a smaller value.

Note that the double Monte Carlo method works for evaluating the power for a single parameter. If one needs to calculate power for more than one parameter, separate analysis has to be conducted. We use two examples to show how to use the method in WebPower.

20.2.1 Example 1. A structural equation model

Suppose in the SEM model in Figure 20.1.3, we are particularly interested in the path from X1 to X2. Then, we can conduct a power analysis based on the double Monte Carlo method for the parameter. In the path diagram, the path or parameter is labeled as a. To request the power analysis, we would need to provide certain information using the input fields on the left panel as shown in Figure 20.2.1.

Note that in the field of "*MC replications*", two numbers are provided. The first is the number of replications (Q) in the first Monte Carlo to get the sampling distribution under the null hypothesis and the second is the number of replications (R) in the second Monte Carlo to get the sampling distribution under the alternative hypothesis. The two numbers are separated by a comma (,). In addition, in the "*Power parameters*" field, we use "test=a" to let WebPower know that the parameter of interest is a. Currently, WebPower can only conduct double Monte Carlo for one parameter at a time.

Example 20.2.1: Power for an SEM model

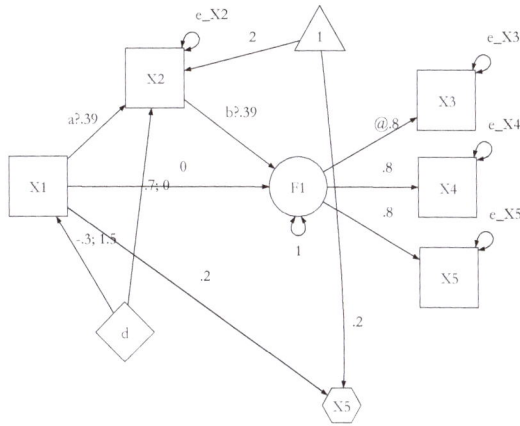

Figure 20.2.1: Power analysis for the parameter a based on double Monte Carlo.

For the model specified in Figure 20.2.1, the output of the power analysis is shown in Figure 20.2.2. Note that on the top of the output, the power for the particular path a is provided. Based on the likelihood ratio test, the power is 0.975. The second part of the output is actually Monte Carlo based power estimates for each parameter in the model. Note that the power from the two methods is slightly different. This is likely due to the method based on the single Monte Carlo has a stronger assumption than the double Monte Carlo method.

20.2.2 Example 2. A two-level model

Similarly, double Monte Carlo based power analysis can be conducted for multilevel models. Figure 20.2.3 shows the input for the two-level model with missing and non-normal data. Again, test=a is specified for testing the single parameter a in the model.

For the model specified in Figure 20.2.3, the power analysis is shown in Figure 20.2.4. The power for the parameter a is given at the bottom of the output, which is 0.405.

Example 20.2.2: Power for a two-level model

Power for parameter a based on simulated critical value
Power 0.974975
s.e. 0.004941983

Power results based on normal Monte Carlo simulation

Figure 20.2.2: Power analysis results for the two-level model with missing data in Figure 20.2.1.

```
Basic information:

   Esimation method                             ML
   Standard error                  robust.huber.white
   Number of requested replications             1000
   Number of successful replications            1000
   Sample size                                   100
```

		True	Estimate	MSE	SD	Power	Power.se	Coverage
Latent variables:								
F1 =~								
X3		0.800	0.800	0.000	0.000	NaN	NaN	1.000
X4		0.800	0.856	0.339	0.350	1.000	0.000	0.954
X5		0.800	0.852	0.527	0.932	1.000	0.000	0.960
Regressions:								
X2 ~								
X1	(a)	0.390	0.396	0.099	0.101	1.000	0.000	0.953
F1 ~								
X2	(b)	0.390	0.398	0.149	0.150	1.000	0.000	0.951
X1		0.000	-0.003	0.147	0.148	0.954	0.007	0.954
Intercepts:								
X2		2.000	2.005	0.099	0.100	1.000	0.000	0.953
X3		0.000	-0.016	0.270	0.275	0.950	0.007	0.950
X4		0.000	-0.012	0.272	0.269	0.962	0.006	0.962
X5		0.000	0.005	0.317	0.328	0.950	0.007	0.950
F1		0.000	0.000	0.000	0.000	NaN	NaN	1.000
Variances:								
X2	(e_X2)	1.000	0.981	0.138	0.142	1.000	0.000	0.947
F1		1.000	1.020	0.457	0.464	1.000	0.000	0.942
X3	(e_X3)	1.000	0.944	0.295	0.301	0.999	0.001	0.961
X4	(e_X4)	1.000	0.946	0.308	0.303	0.997	0.002	0.952
X5	(e_X5)	1.000	0.917	0.518	0.824	0.998	0.001	0.956
Indirect/Mediation effects:								
ab		0.152	0.157	0.072	0.071	1.000	0.000	0.951

Figure 20.2.3: Path diagram for a two-level model with missing data

```
Regular Monte Carlo Power:
  Esimation method                                          REML
  Number of requested replications                          1000
  Number of successful replications                         1000
  Number of levels                                             2
  Sample size for different levels                        100,30
```

Figure 20.2.4: Power analysis results for the two-level model with missing data in Figure 20.2.3.

Level	Equation	Label	True	Estimate	MSE	SD	Power	Power.se	Coverage
	Fixed effects								
2	b0~1		0.3	0.298	0.165	0.162	0.458	0.016	0.942
2	b1~1		0	-0.010	0.174	0.178	0.069	0.008	0.931
2	b0~Z1		0.45	0.448	0.169	0.172	0.744	0.014	0.948
2	b1~Z1	a	0.3	0.305	0.178	0.183	0.435	0.016	0.944
	Random effects								
1	Y1~~Y1	e_Y1	.6	0.600		0.015			
2	b0~~b0	e_b0	.8	0.795		0.208			
2	b1~~b1	e_b1	.9	0.885		0.239			
2	b0~~b1		0	-0.002		0.162			

```
Monte Carlo Power Based on Empirical Distribution of Likelihood Ratio
Parameter    Power
      a      0.405

Monte Carlo Power Based on Empirical Distribution of Parameter Estimation
Parameter    Power
      a      0.393
Replication= 10000
```

21 *Bibliography*

Babin, B. J., Hair, J. F., & Boles, J. S. (2008). Publishing research in marketing journals using structural equation modeling. *The Journal of Marketing Theory and Practice*, 16(4), 279–286.

Blanca, M. J., Arnau, J., López-Montiel, D., Bono, R., & Bendayan, R. (2013). Skewness and kurtosis in real data samples. *Methodology*, 9, 78–84.

Browne, M. W. & Cudeck, R. (1992). Alternative ways of assessing model fit. *Sociological Methods & Research*, 21(2), 230–258.

Cain, M. K., Zhang, Z., & Bergeman, C. (2018). Time and other considerations in mediation design. *Educational and Psychological Measurement*.

Cain, M. K., Zhang, Z., & Yuan, K.-H. (2017). Univariate and multivariate skewness and kurtosis for measuring nonnormality: Prevalence, influence and estimation. *Behavior Research Methods*, 49(5), 1716–1735.

Champely, S. (2012). *pwr: Basic functions for power analysis*. R package version 1.1.1.

Cheung, M. W. L. (2007). Comparison of approaches to constructing confidence intervals for mediating effects using structural equation models. *Structural Equation Modeling*, 14(2), 227–246.

Cohen, J. (1988). *Statistical power analysis for the behavioral sciences*. Hillsdale, NJ: Lawrence Ehrlbaum Associates, 2nd edition.

Cohen, J. (1992). A power primer. *Psychological bulletin*, 112(1), 155–159.

Daly, L. E. (1991). Confidence intervals and sample sizes: don't throw out all your old sample size tables. *BMJ: British Medical Journal*, 302, 333–336.

Deb, P., Trivedi, P. K., et al. (1997). Demand for medical care by the elderly: a finite mixture approach. *Journal of Applied Econometrics*, 12(3), 313–336.

Demidenko, E. (2007). Sample size determination for logistic regression revisited. *Statistics in Medicine*, 26, 3385 – 3397.

DiSantostefano, R. L. & Muller, K. E. (1995). A comparison of power approximations for Satterthwaite's test. *Communications in Statistics – Simulation and Computation*, 24(3), 583–593.

Du, H., Liu, F., & Wang, L. (2017a). A bayesian "fill-in" method for correcting for publication bias in meta-analysis. *Psychological Methods*, 22(4), 799–817.

Du, H. & Wang, L. (2016). A bayesian power analysis procedure considering uncertainty in effect size estimates from a meta-analysis. *Multivariate behavioral research*, 51(5), 589–605.

Du, H., Zhang, Z., & Yuan, K.-H. (2017b). Power analysis for t-test with non-normal data and unequal variances. In L. A. van der Ark, M. Wiberg, S. A. Culpepper, J. A. Douglas, & W.-C. Wang (Eds.), *Quantitative Psychology – The 81st Annual Meeting of the Psychometric Society, Asheville, North Carolina, 2016. Springer Proceedings in Mathematics & Statistics.* (pp. 373–380).: Springer.

Ellis, P. D. (2010). *The essential guide to effect sizes: Statistical power, meta-analysis, and the interpretation of research results.* New York, NY: Cambridge University Press.

Faul, F., Erdfelder, E., Buchner, A., & Lang, A.-G. (2009). Statistical power analyses using g* power 3.1: Tests for correlation and regression analyses. *Behavior research methods*, 41(4), 1149–1160.

Ferraiolo, J., Jun, F., & Jackson, D. (2000). *Scalable vector graphics (SVG) 1.0 specification.* iuniverse.

Fleiss, J. L. (1994). Measures of effect size for categorical data. In H. Cooper & L. V. Hedges (Eds.), *The handbook of research synthesis* (pp. 245–260). New York, NY: Russell Sage Foundation.

Fritz, M. S. & MacKinnon, D. P. (2007). Required sample size to detect the mediated effect. *Psychological Science*, 18, 233–239.

Fritz, M. S., Taylor, A. B., & MacKinnon, D. P. (2012). Explanation of two anomalous results in statistical mediation analysis. *Multivariate Behavioral Research*, 47, 61–87.

Greenhouse, S. W. & Geisser, S. (1959). On methods in the analysis of profile data. *Psychometrika*, 24(2), 95–112.

Grimm, K. J., Ram, N., & Estabrook, R. (2016). *Growth modeling: Structural equation and multilevel modeling approaches.* Guilford Publications.

Hamagami, F. & McArdle, J. J. (2007). Dynamic extensions of latent difference score models. In S. M. Boker & M. J. Wenger (Eds.), *Data analytic techniques for dynamical systems* (pp. 47–86). Mahwah, NJ: Erlbaum.

Hamagami, F., Zhang, Z., & McArdle, J. J. (2010). Bayesian discrete dynamic system by latent difference score structural equations models for multivariate repeated measures data. In S.-M. Chow, E. Ferrer, & F. Hsieh (Eds.), *Statistical methods for modeling human dynamics: An interdisciplinary dialogue* (pp. 319–348). New York, NY: RoutledgeTaylor & Francis Group.

Hayes, A. F. (2013). *Introduction to Mediation, Moderation, and Conditional Process Analysis: A Regression-Based Approach*. New York, NY: Guilford Press.

Hayes, A. F. & Scharkow, M. (2013). The relative trustworthiness of inferential tests of the indirect effect in statistical mediation analysis: Does method really matter? *Psychological Science*, 24, 1918–1927.

Hedges, L. V. & Rhoads, C. (2010). Statistical power analysis in education research. ncser 2010-3006. *National Center for Special Education Research*.

Hoenig, J. M. & Heisey, D. M. (2001). The abuse of power: the pervasive fallacy of power calculations for data analysis. *The American Statistician*, 55(1), 19–24.

Howell, D. C. (2012). *Statistical methods for psychology*. Cengage Learning.

Hoyle, R. H. (1995). *Structural equation modeling: Concepts, issues, and applications*. Sage Publications.

Huynh, H. & Feldt, L. S. (1976). Estimation of the box correction for degrees of freedom from sample data in randomized block and split-plot designs. *Journal of Educational and Behavioral Statistics*, 1(1), 69–82.

Kendall, M., Stuart, A., & Ord, J. (1994). *Distribution Theory*. Edward Arnold Publishers.

Khine, M. S. (2013). *Application of structural equation modeling in educational research and practice*. Springer.

Kline, R. B. (2011). *Principles and Practice of Structural Equation Modeling*. Methodology in The Social Sciences. New York, NY: The Guilford Press, 3 edition.

Liu, X. S. (2013). *Statistical power analysis for the social and behavioral sciences: Basic and advanced techniques*. Routledge.

MacCallum, R. C. & Austin, J. T. (2000). Applications of structural equation modeling in psychological research. *Annual review of psychology*, 51(1), 201–226.

MacCallum, R. C., Browne, M. W., & Sugawara, H. M. (1996). Power analysis and determination of sample size for covariance structure modeling. *Psychological methods*, 1(2), 130–149.

MacKinnon, D. P. (2008). *Introduction to statistical mediation analysis*. New York, NY: Taylor & Francis.

MacKinnon, D. P., Lockwood, C. M., & Williams, J. (2004). Confidence limits for the indirect effect: Distribution of the product and resampling methods. *Multivariate Behavioral Research*, 39(1), 99–128.

Mai, Y. & Zhang, Z. (2017). Statistical power analysis for comparing means with binary or count data based on analogous ANOVA. In L. A. van der Ark, M. Wiberg, S. A. Culpepper, J. A. Douglas, & W.-C. Wang (Eds.), *Quantitative Psychology - The 81st Annual Meeting of the Psychometric Society, Asheville, North Carolina, 2016*: Springer.

Mauchly, J. W. (1940). Significance test for sphericity of a normal n-variate distribution. *The Annals of Mathematical Statistics*, 11(2), 204–209.

Maxwell, S. E. & Cole, D. A. (2007). Bias in cross-sectional analyses of longitudinal mediation. *Psychological Methods*, 12, 23–44.

Maxwell, S. E. & Delaney, H. D. (2003). *Designing Experiments and Analyzing Data: A Model Comparison Perspective*. Routledge.

Maxwell, S. E. & Delaney, H. D. (2004). *Designing experiments and analyzing data: A model comparison perspective*. Mahwah, N.J.: Lawrence Erlbaum Associates, 2 edition.

McArdle, J. J. (2000). A latent difference score approach to longitudinal dynamic structural analyses. In R. Cudeck, S. du Toit, & D. Sôrbom (Eds.), *Structural Equation Modeling: Present and future* (pp. 342–380). Lincolnwood, IL: Scientific Software International.

McArdle, J. J. & Hamagami, F. (2001). Latent difference score structural models for linear dynamic analyses with incomplete longitudinal data. In L. M. Collins & A. G. Sayer (Eds.), *New methods for the analysis of change* (pp. 139–175). Washington, DC: American Psychological Association.

McArdle, J. J. & Nesselroade, J. R. (2014). *Longitudinal data analysis using structural equation models*. American Psychological Association.

Meehl, P. E. (1997). The problem is epistemology, not statistics: Replace significance tests by confidence intervals and quantify accuracy of risky numerical predictions. In L. L. Harlow, S. A. Mulaik, & J. H. Steiger (Eds.), *What if there were no significance tests?* (pp. 393–425). Mahwah, NJ: Lawrence Erlbaum Associates.

Micceri, T. (1989). The unicorn, the normal curve, and other improbable creatures. *Psychological Bulletin*, 105(1), 156–166.

Moore, D. S., Notz, W., & Fligner, M. A. (2013). *Essential statistics*. New York, NY: WH Freeman & Company.

Moser, B. K., Stevens, G. R., & Watts, C. L. (1989). The two-sample t test versus Satterthwaite's approximate F test. *Communications in Statistics-Theory and Methods*, 18(11), 3963–3975.

Muller, K. E., Lavange, L. M., Ramey, S. L., & Ramey, C. T. (1992). Power calculations for general linear multivariate models including repeated measures applications. *Journal of the American Statistical Association*, 87(420), 1209–1226.

Muthén, L. K. & Muthén, B. O. (2002). How to use a monte carlo study to decide on sample size and determine power. *Structural Equation Modeling*, 9(4), 599–620.

Myatt, G. J. & Johnson, W. P. (2011). *Making Sense of Data III: A Practical Guide to Designing Interactive Data Visualizations*. John Wiley & Sons.

Nachtigall, C., Kroehne, U., Funke, F., & Steyer, R. (2003). should we use sem? pros and cons of structural equation modeling. *Methods of Psychological Research Online*, 8(2), 1–22.

O'Brien, R. G. & Kaiser, M. K. (1985). Manova method for analyzing repeated measures designs: an extensive primer. *Psychological bulletin*, 97(2), 316–333.

Peng, C.-Y. J., Long, H., & Abaci, S. (2012). Power analysis software for educational researchers. *The Journal of Experimental Education*, 80(2), 113–136.

Preacher, K. J. & Hayes, A. F. (2004). Spss and sas procedures for estimating indirect effects in simple mediation models. *Behavior Research Methods, Instruments, & Computers*, 36, 717–731.

Raudenbush, S. W. (1997). Statistical analysis and optimal design for cluster randomized trials. *Psychological Methods*, 2(2), 173–185.

Raudenbush, S. W. & Liu, X. (2000). Statistical power and optimal design for multisite randomized trials. *Psychological methods*, 5(2), 199–213.

Rencher, A. C. & Schaalje, G. B. (2008). *Linear Models in Statistics*. Hoboken, N.J: Wiley-Interscience, 2 edition edition.

Resig, J. et al. (2009). *jquery: The write less, do more, javascript library*.

Rosseel, Y. (2012). lavaan: An R package for structural equation modeling. *Journal of Statistical Software*, 48, 1–36.

Satorra, A. & Saris, W. E. (1985). Power of the likelihood ratio test in covariance structure analysis. *Psychometrika*, 50(1), 83–90.

Schmidt, F. L. & Hunter, J. E. (2014). *Methods of meta-analysis: Correcting error and bias in research findings*. Sage publications.

Shadish, W. R., Cook, T. D., & Campbell, D. T. (2002). *Experimental and quasi-experimental designs for generalized causal inference*. Wadsworth Cengage learning.

Shook, C. L., Ketchen, D. J., Hult, G. T. M., & Kacmar, K. M. (2004). An assessment of the use of structural equation modeling in strategic management research. *Strategic management journal*, 25(4), 397–404.

Shrout, P. E. & Bolger, N. (2002). Mediation in experimental and nonexperimental studies: New procedures and recommendations. *Psychological Methods*, 7, 422–445.

Sobel, M. E. (1982). Asymptotic confidence intervals for indirect effects in structural equation models. In S. Leinhardt (Ed.), *Sociological methodology* (pp. 290–312). San Francisco: Jossey-Bass.

StataCorp, L. (2013). Stata power and sample-size reference manual.

Thoemmes, F., MacKinnon, D. P., & Reiser, M. R. (2010). Power analysis for complex mediational designs using monte carlo methods. *Structural Equation Modeling*, 17, 510–534.

Welch, B. L. (1947). The generalization ofstudent's' problem when several different population variances are involved. *Biometrika*, 34(1/2), 28–35.

Westland, J. C. (2015). *Structural Equation Models: From Paths to Networks*, volume 22. Springer.

Wilks, S. S. (1938). The large-sample distribution of the likelihood ratio for testing composite hypotheses. *The Annals of Mathematical Statistics*, 9(1), 60–62.

Yuan, K.-H. & Bentler, P. M. (2006). Mean comparison: Manifest variable versus latent variable. *Psychometrika*, 71(1), 139–159.

Yuan, K.-H. & Maxwell, S. (2005). On the post hoc power in testing mean differences. *Journal of Educational and Behavioral Statistics*, 30(2), 141–167.

Yuan, K.-H., Tong, X., & Zhang, Z. (2015). Bias and efficiency for sem with missing data and auxiliary variables: Two-stage robust method versus two-stage ml. *Structural Equation Modeling: A Multidisciplinary Journal*, 22(2), 178–192.

Yuan, K.-H., Zhang, Z., & Zhao, Y. (2017). Reliable and more powerful methods for power analysis in structural equation modeling. *Structural Equation Modeling: A Multidisciplinary Journal*, 24(3), 315–330.

Zhang, Z. (2014). Monte carlo based statistical power analysis for mediation models: Methods and software. *Behavior research methods*, 46(4), 1184–1198.

Zhang, Z. (2018). Moments of a distribution. In *The SAGE Encyclopedia of Educational Research, Measurement, and Evaluation* (pp. 1084–1085).

Zhang, Z., Hamagami, F., Grimm, K. J., & McArdle, J. J. (2015). Using R package RAMpath for tracing SEM path diagrams and conducting complex longitudinal data analysis. *Structural Equation Modeling: A Multidisciplinary Journal*, 22(1), 132–147.

Zhang, Z. & Liu, H. (2018). Sample size and measurement occasion planning for latent change score models through monte carlo simulation. In E. Ferrer, S. M. Boker, & K. J. Grimm (Eds.), *Advances in Longitudinal Models for Multivariate Psychology: A Festschrift for Jack McArdle*. Taylor & Francis.

Zhang, Z. & Wang, L. (2009). Statistical power analysis for growth curve models using sas. *Behavior Research Methods*, 41, 1083–1094.

Zhang, Z. & Wang, L. (2013). Methods for mediation analysis with missing data. *Psychometrika*, 78, 154–184.

Zu, J. & Yuan, K. H. (2010). Local influence and robust procedures for mediation analysis. *Multivariate Behavioral Research*, 45, 1–44.

22 *Index*